Technological Responses to the Greenhouse Effect

Edited by
GEORGE THURLOW

PhD, C Eng, F I Mech E, F Inst E
Chairman of the Working Group on the Greenhouse Effect appointed by The Watt Committee on Energy

Report Number 23

Published on behalf of
THE WATT COMMITTEE ON ENERGY
by
ELSEVIER APPLIED SCIENCE
LONDON and NEW YORK

ELSEVIER SCIENCE PUBLISHERS LTD
Crown House, Linton Road, Barking, Essex 1G11 8JU, England

Sole Distributor in the USA and Canada
ELSEVIER SCIENCE PUBLISHING CO., INC
655 Avenue of the Americas, New York, NY 10010, USA

WITH 37 TABLES AND 38 ILLUSTRATIONS

©1990 THE WATT COMMITTEE ON ENERGY
Savoy Hill House, Savoy Hill, London WC2R OBU

British Library Cataloguing in Publication Data applied for

ISBN 1-85166-543-9

Library of Congress Cataloging-in-Publication Data applied for

The views expressed in this Report are those of the authors of the papers and contributors to the discussion individually and not necessarily those of their institutions or companies or of The Watt Committee on Energy.

No responsibility is assumed by the Publisher for any injury and/or damage to persons or property as a matter of products liability, negligence or otherwise, or from any use or operation of any methods, products, instructions or ideas contained in the material herein.

Special regulations for readers in the USA

This publication has been registered with the Copyright Clearance Centre Inc. (CCC). Salem, Massachusetts. Information can be obtained from the CCC about conditions under which photocopies of parts of this publication may be made in the USA. All other copyright questions, including photocopying outside the USA, should be referred to the publisher.

All rights reserved. No part of this publication may be reproduced, stored in a retrieval system, or transmitted in any form or by any means, electronic, mechanical photocopying, recording, or otherwise, without the prior written permission of the publisher.

Typeset at the Alden Press Oxford, London and Northampton

Technological Responses to the Greenhouse Effect

Members of The Watt Committee on Energy Working Group on The Greenhouse Effect

This report has been compiled by the Working Group on The Greenhouse Effect
The members of the Working Group were:

G. Barrett, *Environmental Manager, PowerGen*
Dr L. W. Blank, *Environment Department, National Power*
J. H. Boddy
Dr B. C. Bulloch, *Group Utilities Manager, ICI Chemicals & Polymers Ltd.*
Dr N. A. Burdett, *Environmental Policy Development Manager, National Power*
Dr A. G. Crane, *Environment Dept., National Power*
Dr A. Duran – Lopez, *ENDESA*
Dr K. Gregory, *Operational Research Executive, British Coal*
R. S. Hackett
J. Hayes, *Technical Manager, Institute of Petroleum*
G. Henderson, *Building Research Establishment*
Dr T. Hill, *Environmental Scientist, PowerGen*
Prof. P. G. Jarvis, *Institute of Chartered Foresters*
Dr. N. Milbank, *Building Research Establishment*
M. Nomine, *Directeur des Recherches Adjoint, CERCHAR*
Dr M. A. Plint, *Royal Geographical Society*
Dr F. E. Shephard, *Manager, Analytical Division, British Gas*
Prof. C. R. W. Spedding, *Pro-Vice-Chancellor, University of Reading*
Dr G. G. Thurlow, *Chairman*
Dr J. F. Walker, *National Power*
Prof. A. Williams, *Dept of Fuel & Energy, University of Leeds*
Dr M. Williams, *Air Pollution Division, Warren Spring Laboratory*
N. G. Worley, *Deputy Chairman, The Watt Committee*

Correspondent Members of the Working Group

Dr G. E. Angevine, *Canadian Energy Research Institute*
J. Coffey, *Commission of the European Communities*
Dr Ing. B. Delogu, *Directorate-General, Environment, Consumer Protection & Nuclear Safety, Commission of the European Communities*
Dr. G. R. Weber, *Gesamptverband des Deutschen Steinkohlenbergbaus*
Dr Rer. Nat. G. Zimmermeyer, *Gesamptverband des Deutschen Steinkohlenbergbaus*

Sponsors

The Watt Committee on Energy gratefully acknowledges the assistance given by the following organisations which have contributed to the cost of this project:
British Coal
British Gas
CERCHAR (France)
Commission of the European Communities (DG XI – Environment, Consumer Protection & Nuclear Safety)
ENDESA (SPAIN)
Gesamptverband des Deutschen Steinkohlenbergbaus (German Hardcoal Mining Organisation)
National Grid Company
National Power
Nuclear Electric
PowerGen

Foreword

Our study leading to this Report and the Report itself are the fruits of a suggestion made to the Watt Committee Executive by Professor Kenneth Mellanby some years ago. He then drew attention to the environmental issues arising on the energy scene, and especially the 'greenhouse effect', at a time when they had not yet become so prominent in public debate.

In 1990, there seems to be a conference or statement on some aspects of these matters almost daily; although the risks that might arise from global warming had been foreseen long before, the limitations of current scientific knowledge are often all too apparent. For this reason, our working group has not attempted to offer yet another assessment of that issue, crucial as it is, but was asked to concentrate on the actions that might be taken if the responsible organs — international bodies, national governments and other organisations as well as industry and the domestic consumer — decide that actions are necessary. Since we called on those who were best qualified to answer that question — that is, people in both industry and the academic world — it may not be without significance that their answers, as stated in this Report, have much in common with those given from quite different backgrounds.

So that their contribution might be published in time for several major public discussions of the subject, they have had to work to a restrictive timetable (the circumstances are explained in the body of the Report). I am grateful to them for the enthusiasm and dedication which they have contributed to this study, as well as their expert knowledge. I must also thank the many organisations that have contributed information to assist the group; their help is acknowledged in the appropriate places.

My special thanks, on behalf of the Watt Committee, are due to George Thurlow, Chairman of the working group, on whom much of the responsibility for assembling the whole document and coordinating the arguments and conclusions has fallen. This study has also set a precedent in that it has received both financial and technical support from sponsors in several countries and from the Commission of the European Communities; their help too is duly acknowledged elsewhere, but I must say here that without them the project would have been less thorough and productive.

Like other documents produced by working groups appointed by the Watt Committee, this Report on *Technological Responses to the Greenhouse Effect* brought together experts from various relevant disciplines. Increasingly, questions concerning energy are not perceived to be confined by national boundaries, and this is all the more true as the environmental implications of the production and use of energy are widely recognised. It was valuable, therefore, that representatives from other countries played an active part. In this respect, the Report therefore represents the wider aspirations adopted by the Watt Committee in recent years.

Earlier versions of the papers published here were circulated to participants in the Consultative Conference that we ran on 24–25 April 1990. It was held in the historic lecture theatre of the Royal Geographical Society (one of our member institutions), to which also our thanks are due. Since then, the working group has revised the papers extensively, taking account of discussion at the Conference and arriving at a final view. The Report is therefore issued as the considered view of the group as a whole, and I wholeheartedly endorse it, on behalf of the Watt Committee on Energy, in the hope that it may assist all those who are now concerned with the extent of global warming and the need for action.

G. K. C. PARDOE
Chairman, The Watt Committee on Energy

Preface

The Watt Committee on Energy is an independent body and a registered charity. Its members are some sixty British professional institutions, all with a concern for energy and covering a wide range of disciplines. The Watt Committee is, therefore, in the unique position of being able to bring together experts, involved in energy in different ways, to consider the implications of the Greenhouse Effect as these relate to the production and use of energy.

The Executive of the Watt Committee, considering what role it could best play in the general debate on the Greenhouse Effect, unanimously agreed that it had to address more specific objectives than a general review of the current understanding about the build-up and effects of gases in the atmosphere contributing to the Greenhouse Effect. It was decided, therefore, that a Working Group should be set up to consider the technical options open to the United Kingdom to limit the net emission of those gases that principally contribute to the Greenhouse Effect, such as carbon dioxide, methane, CFCs, nitrous oxide and ozone (the so-called 'Greenhouse Gases'). Such technical options will need to be achievable within the next one or two decades and within a sustainable financial burden without leading to further unacceptable environmental or social problems.

While discussion was to concentrate on the situation within the UK, it was thought that this study could be a pilot study, which might then be adapted to assess the situation in other industrialised countries.

It was decided that this Working Group would not be the most appropriate body to discuss the rate and effects of the build-up of global temperature or its dependence on the build-up of the greenhouse gases. These subjects are not, therefore, addressed in the report that follows, other than by a brief summary of the evidence that the concentration of the greenhouse gases in the atmosphere has increased in recent years, necessary to set the scene for the subsequent sections.

What this study does do is to propose a range of measures that is likely to be most effective in bringing about a reduction in the rate of greenhouse gas emissions in the UK in the short term.

The application of this 'basket' of measures, if applied with sufficient determination would, for example, enable the UK to stabilise its contribution to global emissions at 1990 levels by the year 2005. The measures could be taken, if necessary, as a unilateral action as their implementation would not have a major detrimental effect on the economy. To maintain or improve on these levels of emission into the next century will, however, need further measures; and processes are listed which will be needed to meet the demand to reduce emissions once international agreement has been reached on global action. It is proposed that research and development should be started immediately to ensure that these processes are developed in time.

Whereas many of the options – such as reducing the amount of energy used; making more use of 'renewable' energy sources, including biomass and waste materials; banning CFCs; and afforestation — have been discussed extensively elsewhere in recent months, the experts on the working group have studied them as a package. They have expressed their opinions, at least in qualitative terms, as to the reductions that might realistically be achieved, option by option; and over what timescale and at what cost the level of reductions might be achieved. The effects of the application of each option on other factors relevant to the standard of living have also been considered.

The working group has produced this report in somewhat less than a year and is conscious that many of the aspects discussed could, and almost certainly should, be studied in more depth.

It was thought, however, that a study of this depth will be most valuable — even if supplemented by more detailed work later — if it is available in time to contribute to an awareness of the various alternative strategies open to the governments of the UK and other countries, and their desire to take positive actions to curb the emission of the greenhouse gases as the pressure of international debate builds up over the next few years.

The report is divided into four sections.

In the first part, following a brief review of the nature of the 'Greenhouse Effect', there is a discussion of the way in which man-made emissions enhance the amount of the greenhouse gases naturally present in the atmosphere, and their part in the natural exchange of these gases between the atmosphere and biosphere (that part of the earth's surface inhabited by living things) and the oceans. The relative effect of an equivalent mass of the different greenhouse gases (e.g. carbon dioxide as compared with methane or CFCs) on the Greenhouse Effect is considered and the factors that have to be taken into account in comparing the effect of one gas with another are arrived at.

A compilation is presented of all the sources of emission of the various greenhouse gases occurring at the present time in the United Kingdom as far as data are available. These values are related to the total global emissions.

The second section of the report discusses all the major options available to the UK for reducing the release of the greenhouse gases. Clearly, most emphasis is on the production and use of energy, although non-energy sources of these gases are also considered where it is possible to consider altering the amount of release (as with CFCs, and in agriculture and forestry). The potential for the UK to offer financial and technical support to countries that may be much larger (or potentially larger) emitters of the greenhouse gases, and where the technology, being less advanced than in the UK, offers more scope for improvement, is also discussed, as is the case for British support for activities such as reforestation in more tropical regions of the world.

It is clear that the most effective way by which the UK could contribute to the reduction of the greenhouse gases will be by a mix of various options. The third section of the report gives the views of the Working Group as to how this mix should be made up. Opinions on this must be, to some extent, subjective, and there is no one solution, but it is hoped that the views of the Working Group will contribute to and stimulate the informed debate that is necessary before Governmental or other action is taken, particularly if this includes regulation or taxation.

This report is primarily related to the position within the UK; but, as was said earlier, it is hoped that it will also act as a pilot study which other countries might use as a model for similar studies. The Working Group has been greatly helped by the active participation of delegates from companies in France and Spain, and there have been 'correspondent' members from West Germany and Canada. The Environmental Directorate of the Commission of the European Communities has not only sponsored but has also participated in the study. In the final section of the Report, there are commentaries from individuals of the various countries, other than the UK, that have taken part. They place particular emphasis on the differences between the situations in these countries and in the UK.

The Working Party had the opportunity to hear the views of a wide range of delegates at a Consultative Conference held at the Royal Geographical Society, London, on April 24–25 1990. The comments relevant to this report made at the Conference have been noted and taken into account in the final editing of this report. Where specific contributions made either at the Conference or in correspondence to the Working Party are referred to in the text, references to these are given in the list at the end of the appropriate Chapter.

The text of these contributions has been collated into a separate document which can be obtained from the Watt Committee on Energy, Savoy Hill House, Savoy Hill, London WC2R OBU.

The Working Party and the Watt Committee would like to take this opportunity to thank all the many people who have contributed to the work of the Working Party and in particular to those who participated in the Consultative Conference.

G. G. THURLOW
Chairman, Working Group

Contents

Members of The Watt Committee on Energy Working Group on The Greenhouse Effect ii

Foreword v
G. K. C. PARDOE

Preface vii
G. G. THURLOW

Section 1	Introduction to the greenhouse effect	1
Section 2	Assessing the future importance of the main greenhouse gases	5
Section 3	Emissions of greenhouse gases	17
Section 4	Energy conversion and the release of greenhouse gases	27
Section 5	Energy usage in the home, commerce and industry, and its effect on the release of greenhouse gases	35
Section 6	Energy usage in transportation	53
Section 7	Non energy related sources and sinks for the greenhouse gases in the UK	61
Section 8	Scope for UK assistance outside the UK	65
Section 9	Implications for policy formulation	73
Section 10	The view from outside the UK	77
Section 11	Closing remarks	87

The Watt Committee on Energy: Objectives, Historical Background and Current Programme . . 91

Member Institutions of The Watt Committee on Energy 93

Watt Committee Reports 94

Index 95

Section 1

Introduction to the Greenhouse Effect

1.1 INTRODUCTION

A detailed assessment of the science which underpins projections of global warming was not part of the remit set by the Watt Committee for the Technological Responses Working Group. Other reports, in particular the Science Assessment report from the Intergovernmental Panel on Climate Change (WMO/UNEP 1990), are available, both to provide scientific background[a] and to demonstrate the need for a study of technological responses such as that carried out by the Working Group. However, it was considered useful to summarise the key elements of the problem and to describe briefly how these predictions are made.

1.2 THE EARTH'S HEAT BALANCE AND THE GREENHOUSE EFFECT

The first point to note is that there is a natural greenhouse effect which is vital for life on earth. If the absorption of solar radiation were the only process heating the earth's surface, the surface temperature would be well below freezing—about −18°C if the planet's reflectivity remained as at present. The much higher temperature actually sustained—about 15°C on average—arises because certain trace gases in the atmosphere absorb and re-emit a substantial fraction of the infrared radiation which the surface emits in response to solar heating and which in the absence of these gases would escape directly to space. The downward component of this re-emitted radiation warms the surface and lower levels of the atmosphere. This process is termed the greenhouse effect and the gases involved are called greenhouse gases. Water vapour and carbon dioxide are the two most important naturally occurring greenhouse gases. Clouds also exert a greenhouse warming at the surface, but unlike the greenhouse gases, they also reflect solar radiation, thereby reducing the solar heating at the ground. Their net effect on the heat balance at the surface therefore depends on which process dominates, and varies with the type and the altitude of the cloud. In the present climate, the cooling effect dominates (Ramanathan et al. 1989).

Evidence that greenhouse gases operate in the way described is found both in satellite observations of radiation from the earth and atmosphere, and in the observed relationships between the surface temperatures and atmospheric compositions of Venus and Mars. There is no doubt, therefore, that increasing the concentration of greenhouse gases in the atmosphere increases the radiative heating at the surface. There is strong circumstantial evidence also of the role of changes in greenhouse gas concentrations in past climatic changes. Ice core measurements indicate that over the past 160 000 years a close correlation existed between temperature and atmospheric carbon dioxide and methane concentrations as these changed between glacial and interglacial periods (Barnola et al. 1987; Chappelaz et al. 1990). Although changes in the orbital characteristics of the planet are believed to be the primary cause of climate change on this timescale, the magnitude of the temperature changes at the end of the ice age was too large to be explained by this mechanism alone. It is suspected that the increases in carbon dioxide and methane arose biogeochemically in response to warming, but that they then enhanced the warming as a result of their greenhouse effect.

The effect of an increase in the atmospheric concentration of a particular greenhouse gas depends on several factors: its molecular structure, the

---

[a] The IPCC report suggests that on a 'business as usual' scenario, global mean temperature is likely to be about 1°C above the present value by 2025 and about 3°C higher before the end of the next century. The corresponding rise in average sea level is expected to be about 20 cm by 2030 and about 65 cm by 2100. The possibility of such significant changes demonstrates the need for studies of techological responses, such as that carried out by the Working Group.

amount of radiation emanating from the earth in the spectral frequency over which the gas absorbs, the concentration already present of this gas and any other gases which absorb at the same frequency, and, of course, on the size of the increase in concentration.

Although water vapour is the most important natural greenhouse gas, it is important to note that its concentration in the atmosphere is largely determined by the climate itself. Thus water vapour emissions from human activity do not pose a greenhouse threat. The water vapour content of the atmosphere will increase with increasing temperature, however, and this will add to the greenhouse enhancement caused by increases in other greenhouse gases. Of the main greenhouse gases whose concentrations are being increased by human activity, carbon dioxide makes the largest contribution on account of the large amounts emitted compared with the other gases, although per molecule it is the weakest absorber. A more detailed comparison of the effectiveness of the different greenhouse gases is made in Section 2.5 of this report.

1.3 THE CLIMATE RESPONSE

The direct change in the radiative balance induced by an increase in a greenhouse gas can be estimated quite accurately. The main uncertainty arises in evaluating the effect that this has on the climate system. A doubling of the carbon dioxide concentration, for example, would, in the absence of any other climatic changes, lead to a warming of just over 1°C at the surface. Current climate models (see below) suggest that this would be enhanced by anything between 0.5° and 3.5°C as a result of feedback processes.

One of the best understood feedback processes is the water vapour feedback. As mentioned above, increase in temperature would lead to greater evaporation and a rise in the water vapour concentration in the atmosphere which would further enhance the warming. The operation of this feedback in today's climate has been demonstrated by analysis of satellite radiation budget measurements (Raval and Ramanathan 1989). Another positive feedback involves the reduction in snow and ice cover that would occur as a result of greenhouse warming. This would reduce the reflectivity of the surface and allow greater absorption of solar radiation, so enhancing the warming. A potentially very important feedback whose sign has yet to be established will arise from changes in cloud characteristics. Although clouds provide a net cooling effect on the climate, this fact does not, of course, imply that the net effect of any changes in cloud that might accompany global warming will be a cooling. This feedback involves not only changes in the spatial extent and height distribution of clouds, but also such details as changes in their water content, drop size distribution and the proportions of liquid water and ice cloud present (Mitchell et al. 1989).

1.4 CLIMATE MODELLING

It is not appropriate here to review the current projections of climate models. These are adequately documented elsewhere (e.g. WMO/UNEP 1990). The following paragraphs serve merely to highlight some of the important characteristics of the art of modelling for those readers unfamiliar with the subject.

The principal tools for calculating the overall effects on climate of increases in greenhouse gas concentrations are the general circulation models. These were developed from numerical weather forecasting models and are based on the mathematical equations that determine the dynamical and physical behaviour of the atmosphere and ocean. In such models, the principal variables defining the state of the system, e.g. temperature, wind speed and direction, humidity, pressure, etc, are held at the points on a three dimensional grid. The differential equations controlling the system are integrated forward in time at each grid point to predict the state of the system at future times. Starting from a motionless atmosphere, the sun can be 'switched on' and the model run to a point where it has developed a 'quasi-steady' climate in which the broad scale features of the global circulation are readily apparent. Year-to-year variations occur in the model climate but there are no long-term drifts. Since the spatial grid scale is typically about 200 km, many smaller scale processes cannot be represented explicitly. They have to be represented (parametrized) in terms of the variables held at the grid points. The accuracy with which this is done can have a large influence on the model results.

Most projections of future climate so far have been based on so-called equilibrium experiments. In these, the quasi-steady 'equilibrium' climate obtained by running a model with present day greenhouse gas concentrations is compared with that obtained when some constant, elevated level of greenhouse gases (e.g. a doubled carbon dioxide concentration) is used. The differences between the two model climates are interpreted as the changes that would eventually result from the specified change in greenhouse gas levels if the climate system was allowed to reach equilibrium. In principle, such model experiments can predict the likely *magnitude* of a climate change but, by virtue of their conception, can say nothing about the *timing*.

Computer capacities have recently expanded to the point where time-dependent simulations of an evolving climate can be made, in which a scenario of future greenhouse gas concentration trends is used to 'drive' the model. In these simulations it is necessary to model the ocean circulation processes in detail since these have a major bearing on the rate at which the climate system can warm in response to rising greenhouse gas concentrations. It needs to be stressed that the development of these time-dependent models is still at an early stage and significant practical difficulties exist in adequately coupling the atmospheric and oceanic parts of the models. A lack of understanding of, and data to quantify, ocean mixing processes is a further impediment to their rapid development. Nevertheless, some important results are beginning to emerge. Since different regions of the ocean are likely to warm at different rates, depending on the vertical mixing patterns, the evolving regional patterns of climate change derived from time-dependent simulations differ in some important respects from equilibrium simulations (e.g. Stouffer et al. 1989). This clearly has consequences for the identification of projected climatic changes in observational records. In addition to making time-dependent climate projections, fully coupled atmosphere–ocean circulation models will also be needed to improve projections of sea-level rise and the carbon uptake capacity of the oceans (see Chapter 2), which at present are largely based on simpler diffusion models.

1.5 OBSERVED CLIMATE CHANGE

A key question often asked concerns whether or not the enhanced greenhouse effect has been observed. There is little doubt that the global mean temperature has risen over the last 100 years by about 0·3–0·6°C (WMO/UNEP 1990). The range of uncertainty takes into account such factors as changing observational practices, increasing urbanisation in the environs of observing stations and the changing density of the observational network. The warming has been accompanied by a retreat of most mountain glaciers (Grove 1988), and a rise in sea level of 10–20 cm (WMO/UNEP 1990). It is not possible to attribute these occurrences unequivocally to an enhanced greenhouse effect, however, since the magnitude of the warming is within the range of natural variability and it has not followed a trend that is consistent with increased greenhouse gas concentrations as the major cause.

Several interpretations are possible, however, none of which can yet be proved or disproved. One is that the overall increase in temperature is due to greenhouse gas increases superimposed on natural variations, in which case the sensitivity of the climate to greenhouse gas increases is in the lower part of the range currently predicted by climate models. Alternatively, it is possible that a large part of the warming has been caused naturally, and that the greenhouse effect is yet smaller. On the other hand, natural or other man-made effects might have masked an otherwise larger warming due to the greenhouse effect, in which case a climate sensitivity in the upper part of the currently projected range might still be possible.

It seems unlikely that a definitive conclusion will be forthcoming before the year 2000, given the size of the natural variability in climate. As the precision with which regional climate changes can be projected improves, however, it will be possible to search more purposefully for patterns and combinations of observed regional climatic changes that together provide evidence that the enhanced greenhouse effect is, or is not, altering global climate as predicted.

REFERENCES

BARNOLA, J. M., RAYNAUD, D., KOROTKEVITCH, Y. S. & LORIUS, C. (1987) *Vostok ice core: a 160,000 year record of atmospheric CO_2*. Nature, 329, 408–14.

CHAPPELLAZ, J., BARNOLA, J. M., RAYNAUD, D., KOROTKEVITCH, Y. S. & LORIUS, C. (1990). Ice-cave record of atmospheric methane over the past 160 000 years. Nature, 345, 127–31.

GROVE, J. M. (1988) *The Little Ice Age*. Methuen, London, 498 pp.

MITCHELL, J. F. B., SENIOR, C. A. & INGRAM, W. J. (1989) CO_2 *and climate: A missing feedback?* Nature, 341, 132–4.

RAMANATHAN, V., CESS, R. D., HARRISON, E. F., MINNIS, P., BARKSTROM, B. R., AHMED, E. & HARTMANN, D. (1989) *Cloud-radiative forcing and climate: results from the earth radiation budget experiment*. Science, 243, 57–63.

RAVAL, A. & RAMANATHAN, V. (1989) *Observational determination of the greenhouse effect*. Nature, 342, 758–61.

STOUFFER, R. J., MANABE, S. & BRYAN, K. (1989) *Interhemispheric asymmetry in climate response to a gradual increase of CO_2*. Nature, 342, 660–2.

WMO/UNEP (1990) *Scientific Assessment of Climatic Change*. Working Group I Report of the Intergovernmental Panel on Climate Change, WMO, Geneva.

Section 2

Assessing the Future Importance of the Main Greenhouse Gases

2.1 INTRODUCTION

An important difference between gaseous emissions associated with acid deposition—sulphur dioxide and nitrogen oxides—and the greenhouse gases is in their rate of cycling in the earth-atmosphere system. In the case of the acid precursor pollutants, atmospheric lifetimes are short with annual emissions and removal many times larger than the amount present in the atmosphere at any time. The effect of man-made emissions of these gases is to enhance the amount of pollutant cycled, rather than to generate a sustained build-up in atmospheric concentration. In contrast, many of the greenhouse gases have very slow rates of removal from the atmosphere and consequently long atmospheric residence times. Man-made emissions, which may be quite small in comparison with the atmospheric abundance, thus have the potential to accumulate.

Atmospheric removal mechanisms and the exchange processes between the atmosphere, oceans and biosphere which control atmospheric concentrations of greenhouse gases are not well understood. This adds to the already large uncertainty in predicting future concentrations that arises from uncertainty in future emission rates. This chapter considers the issues surrounding past and projected future trends in global emissions and atmospheric concentration for each of the main greenhouse gases, and also assesses their relative importance.

2.2 CARBON DIOXIDE

2.2.1 Sources of Emissions

Natural exchanges of carbon dioxide take place between atmosphere and biosphere and between atmosphere and ocean. Exchange rates are largely controlled by prevailing long-term climatic conditions. During periods of stable climate these exchanges remain in approximate balance. A net emission of CO_2 to the atmosphere occurs when materials containing carbon are burned. The combustion of fossil fuels is currently the dominant man-made source of CO_2 amounting to 5·3 billion tonnes of carbon. (Note that this is equivalent to 18·6 billion tonnes of carbon dioxide. In order to compare transfers in the carbon cycle it is customary to quantify CO_2 release in terms of the carbon content.)

Of the fossil fuels, combustion of coal produces the largest CO_2 emission per unit of energy produced and natural gas the least, the ratio coal:oil:gas being approximately 5:4:3. A breakdown of current emissions by fuel and by nation is given in Chapter 3.

The period of global economic development and expanding Third World population over the last four decades has witnessed an increasing shift in the regional sources of fossil fuel CO_2 away from the West towards Eastern Europe, the USSR and the developing countries. Whereas in 1950 North America and Western Europe accounted for 68% of global CO_2 emissions, their contribution had fallen to 43% by 1980. This changing emission pattern has major implications for future policy considerations about emission control.

CO_2 is also released from soils when virgin land is ploughed for agricultural purposes. This release probably made a substantial contribution to atmospheric CO_2 increase during the 19th and early 20th centuries. Much attention is now focussed on the release of carbon dioxide to the atmosphere as a result of deforestation in the tropics. Calculations of its magnitude are seriously hampered by uncertainty in the rate of clearing, the carbon content per unit area of the forests involved and the fraction of this carbon released to the atmosphere. At least part

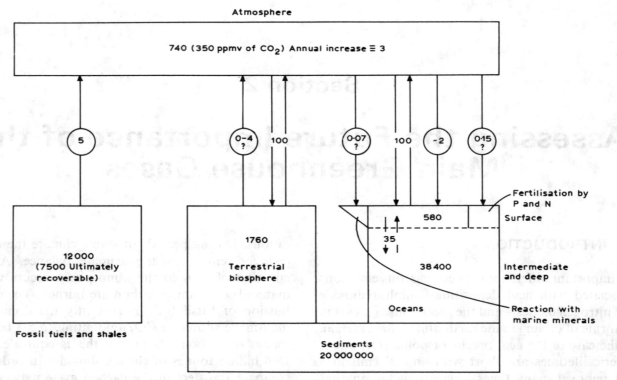

Fig. 2.1. The global carbon cycle. Figures within boxes are total carbon contents (billion tonnes of carbon); figures in double arrows are natural fluxes (billion tonnes C per year); circled figures on single arrows are man-induced fluxes. (After Liss & Crane, 1983.)

of the release from tropical deforestation is compensated by accumulation of carbon in temperate forests, which have expanded following earlier clearance of virgin forest. The global net release of carbon to the atmosphere from deforestation and land-use change has been estimated at 1.8 ± 0.9 billion tonnes, representing about a quarter of the total man-made carbon release. A more detailed account of the role of forests in the carbon cycle is given in Chapter 8.

2.2.2 Carbon Cycling

On timescales of interest to the CO_2/climate change issue (years to centuries) the main natural exchanges of carbon in the earth-atmosphere system are those between the atmosphere and the surface layers of the oceans and between the atmosphere and the biosphere. An illustration of the main reservoirs and transfers of carbon in the global cycle is given in Fig. 2.1. It will be seen that the atmosphere, surface ocean and terrestrial biosphere contain similar amounts of carbon. Soils contain about twice this amount, while the intermediate and deep ocean waters have a dissolved carbon content 50–60 times that in the atmosphere. It is apparent that the oceans will provide the major capacity for the uptake of man-made CO_2. However, the slow rate of exchange between the surface and deeper layers means that the greater part of this capacity will only be utilised very slowly.

The most reliable measurements of atmospheric CO_2 concentration in the period up to the late 1950s have come from analysis of CO_2 in air bubbles trapped in polar ice (Fig. 2.2 inset). These indicate that the natural CO_2 concentration in preindustrial times was about 280 parts per million by volume (ppmv). Since 1958 monitoring stations have been established across the globe providing measurements of the type depicted in Fig. 2.2. The present global mean concentration is close to 350 ppmv. An annual cycle due to the cycling of CO_2 by the biosphere is evident at all sites, but is stronger in higher Northern latitudes where the effects of relatively large biomass and strong climatic seasonality combine. There is some evidence that the amplitude of the annual cycle has increased in the last decade. This may indicate shifts in the seasonality of the uptake and release of atmospheric CO_2 or an increase in biospheric uptake due to higher CO_2 concentrations or higher global temperature.

The rising secular trend is also evident at all sites. The increase in the southern hemisphere lags behind that in the northern hemisphere by a few ppmv. This is because the major sources of man-made CO_2

are in the northern hemisphere and inter-hemispheric transport times are of the order of a year. Mixing within hemispheres is more rapid. Shorter timescale local variations are evident due to plant growth, differing airmass trajectories etc.

The increase in atmospheric CO_2 concentration over time is observed to be significantly less than the man-made input. The Meeting of Experts at the WMO Villach Conference in 1985 (Bolin et al. 1986) concluded that the best estimate of the fraction of CO_2 emissions that have remained airborne is 39% (34–46%) for the period 1958–82 and 43% (32–60%) for the period 1860–1980. These figures are based on the observed atmospheric CO_2 concentration change and estimated biospheric and fossil fuel CO_2 emissions over these periods. Much higher values for the airborne fraction, around 60%, are derived on the basis of carbon cycle models, largely because the oceans do not appear to have been capable of taking up more than about 30–40% of fossil fuel CO_2 emissions. To be consistent with observed CO_2 concentration trends, these models would require there to have been little or no net biospheric release of carbon in recent decades or that other sinks of CO_2 exist which are not account-

ed for in the models. Examples of such possible sinks include increased sedimentation of organic carbon in coastal regions, and a larger fraction of burned biomass being locked up in charcoal than is normally assumed when calculating estimates of biospheric CO_2 release. A recent observational study (Tans et al. 1990), based on atmospheric and oceanic carbon dioxide measurements, has also demonstrated the impossibility of balancing the global carbon budget on the basis of present estimates of known CO_2 sources and sinks. The study suggests that a large, but unidentified, sink for CO_2 must exist on the continents in terrestrial ecosystems if the observed distribution of dissolved CO_2 in ocean surface waters is to be consistent with the observed north–south gradient of CO_2 in the atmosphere.

Over and above the uncertainty that exists in the present value of the airborne fraction, the fraction is likely to change in the future. Generally, the more rapid the emission growth rate, the higher becomes the airborne fraction. Conversely, a slowing of the growth rate would lead to a greater fraction of CO_2 emission being taken up by the oceans. As a result of the changing chemistry of the surface waters in response to CO_2 increase, however, there will be a

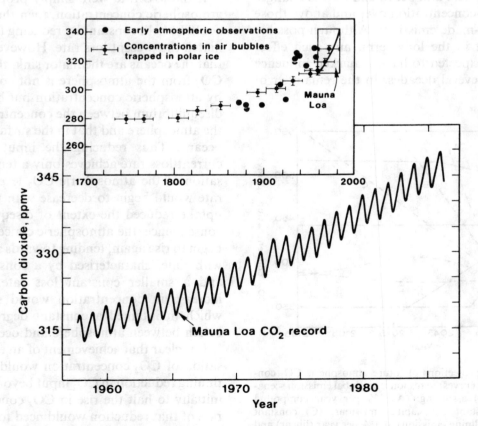

Fig. 2.2. Atmospheric carbon dioxide concentration measured at Mauna Loa, Hawaii since 1958 and (inset) concentrations derived from earlier atmospheric and ice-core data.

tendency for a reduction in CO_2 absorption capacity as CO_2 concentrations rise. In the absence of other effects this will lead to a slow increase in airborne fraction.

2.2.3 Atmospheric Response to Changing Emissions

Although uncertainty in the airborne fraction introduces a significant quantitative uncertainty into projections of future CO_2 concentrations, the main determinant of atmospheric CO_2 concentrations in the next century is likely to be the rate of growth of man-made emissions. It is possible that future atmospheric concentration may also be influenced by climatic feedback processes. The present, natural, atmospheric CO_2 concentration is strongly affected by the biological activity in the upper ocean. This maintains a depletion in the surface water CO_2 concentration relative to that in deeper waters. Climatically-induced changes in biological productivity through, for example, circulation-induced changes in nutrient supply or physical processes affecting the life cycles of phytoplankton, would have the potential to alter the balance of CO_2 between atmosphere and ocean and lead to changes in atmospheric concentration over and above those caused by man-made emissions. Although possibly very significant in the long term, any such effect would not be expected to have a marked influence over the next several decades. In the remainder of this section, therefore, attention is confined to the effects of fossil fuel CO_2 emissions.

There has been considerable discussion in political circles on the concept of stabilising atmospheric CO_2 concentration levels at some point in the future and the emission policies that would be required to achieve this goal. The latter clearly depend on when and at what level stabilisation is required. Carbon cycle models have been used to project future concentration trends for given input scenarios and allow policy questions of this type to be explored. It is beyond the scope of this report to examine these in detail. However, in view of the likely importance of the stabilisation concept in future policy making, it is useful to review some of the fundamental characteristics of the response of the atmospheric CO_2 concentration to changes in emissions since these are not entirely intuitive.

At present, the concentration of CO_2 in the atmosphere is not in equilibrium with the major natural sources and sinks, but is in excess as a result of past man-made emissions. The annual input of man-made emissions currently exceeds the loss of excess CO_2 from the atmosphere, so the concentration is rising.

If the loss rate were simply proportional to the atmospheric concentration, then the concentration could be held constant by reducing the input rate to match the present loss rate. However, to the extent that the oceans are the major sink, the loss of excess CO_2 from the atmosphere is not controlled simply by atmospheric concentration but by the extent of disequilibrium between the concentration of CO_2 in the atmosphere and that in the surface waters of the oceans. Thus reducing the input to match the current loss rate achieves only a temporary stabilisation of the atmospheric CO_2 level since the loss rate would begin to decrease with time as oceanic uptake reduced the extent of disequilibrium. As a consequence, the atmospheric concentration would begin to rise again, tending towards a linear increase with time, characterised by a constant input rate and a smaller constant loss rate. (The rate of increase in concentration would simply be that which maintained a constant degree of disequilibrium between atmosphere and ocean.)

It is clear that achievement of an enduring stabilisation of CO_2 concentration would require a continuing reduction in CO_2 input beyond that required initially to halt the rise in CO_2 concentration. The rate of that reduction would need to match the rate at which the atmosphere-ocean transfer declined. Assuming other factors did not come into play, an

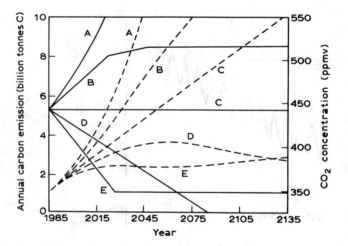

Fig. 2.3. Model projections of future atmospheric CO_2 concentration (dashed curves) in response to global emission scenarios (solid curves) assuming: (A) 2% per year compound growth, (B) constant per capita emissions, (C) constant emissions, (D) declining emissions at 1% per year (linear) and (E) declining emissions at 2% per year (linear) followed by constant emissions. (After Krause et al. 1989.)

indefinite stabilisation would require man-made emissions eventually to fall to zero, at which point a new steady state would be reached, with the ocean at equilibrium with the stabilised atmospheric concentration.

Given that the annual loss of excess CO_2 is about 50–60% of the current man-made CO_2 input, stabilisation at today's atmospheric concentration would require an initial cut in emissions of 40–50%. Since carbon cycle models indicate that excess CO_2 in the atmosphere resulting from a pulse input has a quasi-exponential decay time of 100–200 years (Fig. 2.6), the timescale over which the subsequent reduction to zero input would need to be implemented would be a few hundred years.

It is important to contrast the concept of stabilizing CO_2 concentration with that of stabilizing emission rates. It follows from the above that holding emissions at their present rate would not halt the rise in concentration. It would, however, slow the rate of growth of atmospheric concentration from today's quasi-exponential rate towards a linear growth.

To summarize, therefore, the manner in which the CO_2 emission rate varies over the coming decades will determine whether CO_2 concentration in the atmosphere increases rapidly, slowly or insignificantly. Figure 2.3, based on the results of Krause et al. (1989), plots trends in CO_2 concentration for five input scenarios and illustrates the response characteristics described above.

2.2.4 Projections of Future CO_2 Emissions

Over the past decade many projections have been made of future global CO_2 emissions based on a wide variety of assumptions about future economic, social and technological developments in the developed and the developing world. It is outside the remit of this report to discuss these in any detail. Nevertheless, a brief look at the two principal factors which will have the greatest impact on future energy use and hence on CO_2 emissions, namely population growth and aspirations for economic development, enables some indication of possible future emission rates to be explored.

Table 2.1, based on data published by the World Commission on Environment and Development (1987), shows that in 1984 the high income quarter of the world population consumed three quarters of world energy at an average rate of 6·6 kW per capita. The low income half of the world population consumed about 10% of world energy at an average rate of 0·4 kW. The remaining quarter, classified as middle income, consumed energy at an average rate of 1·3 kW per capita.

Table 2.2 illustrates projected total energy requirements in 2025 for three different development scenarios. In the breakdown of the global population, which is assumed to reach 8·2 billion by then, the growth of the high income group to 1·6 billion includes some transfer from the middle income group of 1984 as well as a very modest increase in the population previously classified as high income. The low and middle income groups are now combined for ease of presentation.

If energy per capita remains at 1984 levels for all groups total world energy use increases by 37% by 2025. If high income energy use per capita remains

Table 2.1 World population and energy consumption according to economic status in 1984

Income	High	Middle	Low
Population	1 141 M	1 188 M	2 390 M
Total Energy	7·58 TW	1·27 TW	0·99 TW
Energy/Capita	6·64 kW	1·07 kW	0·41 kW

Source: World Commission on Environment and Development (1987).

Table 2.2 Projected energy consumption in 2025 for various energy per capita assumptions

	High Income	Low/Mid Income	Total
Population	1 600 M	6 600 M	8 200 M
Energy/Capita		Total Energy	
As 1984	9·3 TW	4·2 TW	13·5 TW
High: as 1984 Med: 2 × 1984 Low: 3 × 1984	9·3 TW	10·5 TW	19·8 TW
High: as 1984 Med: 0·5 × high Low: 0·5 × high	9·3 TW	21·9 TW	31·2 TW
1984 Reference	7·6 TW	2·3 TW	9·8 TW

Entries are total energy consumptions for the high and low + middle income groups for different energy/capita assumptions shown on the left. These assumptions are: As 1984—all population use same energy/capita as 1984. Hence the increase over the 1984 reference (given in the lowest line) reflects simply the rise in population of each group.
The Second entry assumes that by 2025 the high income group have the same energy/capita as in 1984, the middle income group twice their energy/capita in 1984 and the low income group three times their energy/capita in 1984.
The third entry assumes the medium and low income groups increase their energy/capita to half the level enjoyed by the high income group, whose energy/capita remains as in 1984.

as in 1984, but the medium income group doubles its use to just over 2 kW and the low income group triples its energy use to 1·2 kW, a doubling in global energy requirement occurs by 2025. This corresponds with an annual growth rate of 1·7%. It is worth noting that world CO_2 emissions rose at over 4% per annum over most of the period from 1860 to 1973 and have been increasing at well over 2% per annum in the latter half of the 1980's. Goldemberg et al.'s (1985) scenario, based on a global population of 7 billion and a world energy consumption of 11·2 TW by 2020—close to stabilisation at current levels—would require a halving of the energy per capita in the industrialised world to 3·1 kW if the developing countries raised their consumption from 1·0 kW in 1980 to 1·3 kW in 2020. (Note that the disaggregation of world population into economic groups differs between the two studies, but this does not affect the broad message contained in the figures.)

The figures given in the preceding paragraphs indicate that if population does increase along projected lines and fossil fuels remain the world's major energy source then significant increases in CO_2 emissions over the next few decades must be anticipated even when modest assumptions are made about future energy consumption rates.

2.3 METHANE

2.3.1 Methane Emissions

The atmospheric concentration of methane (CH_4) appears to have doubled over the past few hundred years. The trend towards higher concentrations correlates closely with rising human population and it seems likely that it is at least partially man-induced. Expansion of cattle production and rice growing are probably the major man-made sources, but biomass burning, natural gas extraction and coal mining may make significant contributions. The global emission rate is estimated by Bolin et al. (1986) to have been around 275 million tonnes per year in 1940, since when it appears to have approximately doubled. Quantitative estimates of the contributions to the present global and UK sources are tabulated in Chapter 3. It should be noted, however, that both man-made sources and the natural sources and sinks of methane are much less well established than those of CO_2.

2.3.2 The Global Methane Budget

The present atmospheric concentration of methane is about 1·7 ppmv, having risen from a value of around 0·7 ppmv several hundred years ago. Ice core data suggest that the value had changed little in the preceding several thousand years. The present abundance is roughly ten times the annual emission rate, indicating an atmospheric lifetime of about 10 years. The make-up of the total emission is subject to considerable uncertainty. Individual source emission rates reported in the literature differ by more than an order of magnitude. Nevertheless, attempts to assess the methane budget suggest that to explain the observed rise in atmospheric concentration of around 1% per year, sources must currently exceed sinks for methane by some 50 million tonnes per year, i.e. by 10–15%. The most important sink for methane is photochemical oxidation by hydroxyl (OH) radicals in the troposphere. A small fraction reaches the stratosphere before being oxidised. The tropospheric oxidation pathway is influenced by the presence of other species which compete for reaction with OH radicals. It is possible that some of the increase in methane concentration may have been due to a decline of OH concentrations.

As with CO_2 the possibility exists that climatic warming will influence the natural budget of methane. Within near-shore ocean sediments and also in Arctic permafrost, methane exists in hydrated form, known as clathrates. These are stable at the temperature and pressures at which they currently exist, but could be destabilised, so releasing methane, under warmer conditions. Some estimates suggest that the release of methane from clathrates could become comparable with today's total methane emissions, but the extent of the potentially unstable clathrates is still largely a matter of speculation (Lashoff 1989).

2.3.3 Future Methane Concentrations

In view of the uncertainties in current methane emissions, together with the possibility of climatically induced changes in the natural components of the methane budget, it is difficult to project future atmospheric concentrations. As indicated earlier, the concentration will be influenced by any changes in the lifetime of methane in the atmosphere which is largely controlled by the hydroxyl (OH) radical. While future increases in the emissions of methane and carbon monoxide would tend to reduce OH concentrations, thereby increasing methane's lifetime, increased ozone and water vapour concentrations (the latter resulting from climatic warming) could raise OH concentrations and thus have the opposite effect. Over the next few decades a continuation of the current concentration growth rate of about 1% per year is probably the best projection that can be made.

2.4 OTHER GREENHOUSE GASES

2.4.1 Nitrous Oxide

Knowledge of the natural sources and sinks of nitrous oxide (N_2O) is sparse, but it seems likely that man's activities have contributed to the slow increase that has been observed over the past century.

Ice-core data suggest that the pre-industrial concentration was around 0·280 ppmv. In the mid 1980s it had risen to 0·310 ppmv, with the current rate of increase approaching 0·3% per year.

A small proportion of oxidised nitrogen released during fossil fuel combustion is in the form of N_2O. Until very recently measurements were suggesting that this was a significant man-made source. However, sampling defects were discovered to have affected the data and it is now regarded as only a minor contributor. Use of fertilizers, land-use changes and biomass burning are probably the major man-made sources, although these are poorly quantified. An estimated breakdown of natural and man-made sources is given in Chapter 3, showing that the man-made contribution is about one seventh of the present total of 20 million tonnes per year. N_2O is inert in the troposphere and thus has a long lifetime—about 160 years. Its main destruction process is photolysis in the stratosphere, where it is the major source of nitrogen species.

There is no reason to suppose that future emission rates will change significantly from those at present, although it has been suggested that fluidised-bed combustors and catalyst-equipped vehicles may produce substantially higher N_2O emissions than the corresponding conventional technology.

2.4.2 Tropospheric Ozone

Unlike the other major greenhouse gases, ozone is not sufficiently long-lived to exhibit an even distribution in space and time across the globe. It is thus difficult to establish trends in concentration. Measurements made at several locations at the end of the 19th Century indicate that background levels may have doubled. Measurements over recent decades in Europe, USA and Japan have also shown increasing trends, more especially in summer.

Ozone is not emitted to the atmosphere, but created in it. It is widely supposed that the observed increases have resulted from increased emissions of nitrogen oxides (NO_x) and hydrocarbons in industrial and urban regions. However, it has been suggested that climatic changes could also have played a role if there has been an increased frequency of meteorological situations conducive to ozone generation (Davies et al. 1987). Some doubt has also been cast recently on the validity of the observed increase at one key European station following the discovery of possible measurement errors due to local SO_2 pollution (Low et al. 1990).

Estimating the greenhouse effect resulting from tropospheric ozone increase is further complicated by the non-uniform vertical distribution of any changes and sparsity in the global data coverage. Ozone's lifetime in the troposphere is about two months. Thus, to the extent that its concentration is determined by precursor pollutant emissions, it would respond rapidly to appropriate emission reductions. As yet, however, it is not clear what emission control policies would be most effective in reducing tropospheric ozone concentrations.

2.4.3 Chlorofluorocarbons

Chlorofluorocarbons (CFCs) are man-made chemicals not naturally present in the atmosphere. The two most important CFCs of relevance to the greenhouse issue are CFC11 and CFC12. These are used as aerosol propellants (CFC11 and 12), in foam blowing (CFC11), as refrigerants (CFC12) and in air conditioning systems (CFC11). The concentrations in the atmosphere of CFC11 and CFC12 had, by 1988, reached 0·26 and 0·44 parts per billion by volume (ppbv), and are rising at 4% per year. With photolytic destruction in the stratosphere the only significant sink, these gases have long atmospheric lifetimes—about 60–80 years for CFC11 and 90–140 years for CFC12.

Primarily because of their stratospheric ozone depleting properties, the production of CFC11, CFC12 and a range of other CFCs and chlorine containing gases, is being reduced under the terms of the Montreal Protocol. Wigley (1988) has projected future concentrations of CFC11 and 12 in accordance with the protocol and for various degrees of control beyond it (Fig. 2.4). He showed that unless the protocol is significantly streng-

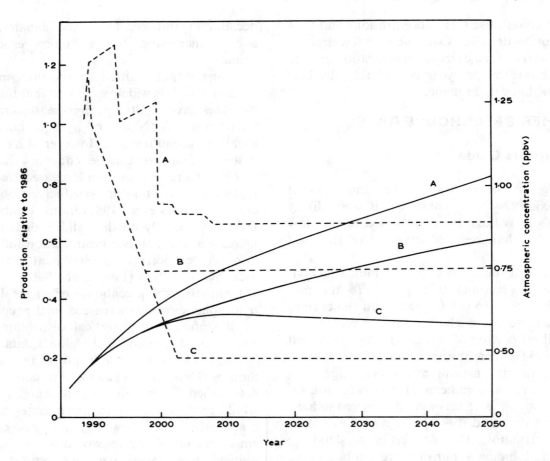

Fig. 2.4. Projected CFC12 atmospheric concentration trends (solid curves) in response to three possible CFC12 production scenarios, expressed as a fraction of 1986 production, reflecting different degrees of compliance with and extension beyond the terms of the Montreal CFC reduction protocol. (Similar trends apply to CFC11).

thened, concentrations of CFC11 and 12 would be expected to double by around 2030.

Replacement compounds for CFCs are being developed in which hydrogen replaces some or all of the chlorine in order to reduce their ozone-depleting potential. Many of the proposed replacements have infra-red absorption strengths comparable with CFCs. However, the presence of hydrogen significantly reduces the lifetime of these gases and thus reduces their greenhouse effectiveness in the longer term relative to CFCs. However, unrestrained emissions of these replacements could still lead to significant contributions to global warming (Shine 1990).

2.5 RELATIVE IMPORTANCE OF THE EMISSIONS OF DIFFERENT GREENHOUSE GASES

The infra-red absorption strength of the different greenhouse gases varies greatly, depending on their molecular structure, the wavelength at which they absorb and the amount of the gas already present in the atmosphere. Table 2.3 shows the warming potential of the main greenhouse gases relative to CO_2 on both a per molecule (molar) and per unit mass basis.

These values, which are still subject to revision as calculations are refined, have been used to rank the importance of the observed increases in greenhouse gases in the atmosphere over different time periods, as illustrated in Fig. 2.5.

When comparing the greenhouse effect of the emissions of different gases into the atmosphere, however, the values in Table 2.3 must be modified to take account of the different lifetimes of the gases in the atmosphere. These vary considerably, as shown in Fig. 2.6, where the fraction of an input remaining in the atmosphere, following an input at zero time, is plotted as a function of time.

Since CO_2, N_2O and CFCs are all long-lived gases, the differences in their decay characteristics does not have a large qualitative impact in modifying their relative effectiveness on a several decades-

Assessing the future importance of the main greenhouse gases

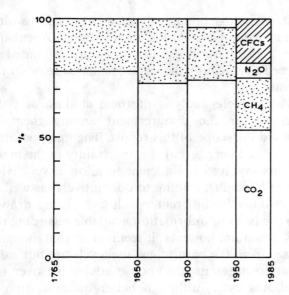

Fig. 2.5. Relative contributions to the greenhouse effect from observed increases in atmospheric concentration of CO_2, methane, nitrous oxide and CFCs over different time periods. (After Wigley 1989.)

Table 2.3 Relative greenhouse effect, per molecule and per unit mass in the atmosphere, of the main greenhouse gases

	Per Molecule	Per Unit Mass
carbon dioxide	1	1
methane	27	74
nitrous oxide	165	165
CFC 11	14 900	4 770
CFC 12	17 700	6 474

century timescale. With its much shorter lifetime, however, methane's effect declines quite rapidly compared with the other gases. The relative effectiveness of a methane emission is thus very sensitive to the time period over which its effect is being considered.

A rough indication of the effect of the different atmospheric lifetimes of CO_2 and methane on the greenhouse effectiveness of their emissions can be deduced from a comparison of observed trends in their atmospheric concentrations. At present methane concentration is increasing at 1% per year, i.e. by 50 million tonnes CH_4 per year. Assuming that natural emissions are in balance with natural sinks and do not contribute to this increase, the rise in concentration is a response to a man-made input of about 360 million tonnes CH_4 per year. This implies an airborne retention (airborne fraction) of 14%. This is approximately one third of the CO_2 airborne fraction discussed earlier. Although airborne fraction is a function of past emission rates and will vary with time, the one third ratio calculated is probably a valid factor by which to modify the concentration effectiveness ratio of methane to CO_2 given in Table 2.3 when considering emissions over periods of a few decades. This implies, therefore, that the effect of methane emissions on this timescale is about 10 times that of CO_2 on a molar basis or 25 times on a mass basis.

A further complication arises with methane in that the chemical reactions involved in its destruction give rise to ozone and stratospheric water vapour formation which enhance the greenhouse effect. Although highly uncertain, some estimates suggest that this indirect greenhouse effect could result in up to a doubling of methane's effectiveness as a greenhouse gas. This factor would also be

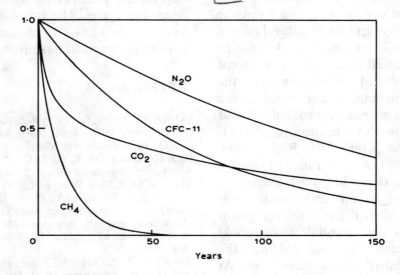

Fig. 2.6. Fraction of emissions of different greenhouse gases remaining in the atmosphere as a function of time following an input at zero time (Rodhe 1989).

strongly time-dependent. If taken into account, the CH_4/CO_2 emission effectiveness ratio would then be 10–20 on a molar basis or 25–50 on a mass basis, with larger values applying in the short term, and smaller values in the long term. These ratios are consistent with recent modelling studies in which the global warming potentials of unit mass emissions of different greenhouse gases have been evaluated (e.g. Derwent 1990, Wilson 1989).

A comparison of today's actual emissions of CO_2 and methane can be made, assuming that the relative effectiveness ratio of 25–50 (mass basis) derived above is appropriate for comparing the average effect of today's emissions over the following several decades. Taking total man-made emissions of CO_2 from fossil fuel combustion and deforestation as 26 billion tonnes of CO_2 (7·1 billion tonnes carbon) and man-made emissions of methane as 365 million tonnes (Table 3.2) it follows that, globally, today's anthropogenic methane emissions will, over the next few decades, have made a contribution which is 35–70% that of today's CO_2 emissions. For the UK, the emission figures given in Chapter 3 (Tables 3.6 and 3.8) indicate a smaller contribution from methane relative to CO_2, namely 13–27%. As already noted, the importance of today's methane emissions relative to CO_2 reduces progressively as longer timespans are considered.

The relative importance of CO_2 and methane emissions is of particular practical importance in comparing the contribution to the greenhouse effect of different fossil fuel cycles, in particular coal and natural gas. While the combustion CO_2 emissions can be accurately evaluated, quantification of the methane releases from coal mining and from the extraction, transportation and use of natural gas are subject to considerable uncertainty. A specific question that is addressed in a number of recent published and unpublished reports concerns the extent to which these methane releases influence the advantage that natural gas has over coal in terms of its reduced combustion CO_2 emission (e.g. Abrahamson 1989, Okken & Kram 1989, Rodhe 1989, Wilson 1989, Eyre 1990). A wide range of conclusions has been drawn, depending on the assumptions made concerning the timescales under consideration, the extent of methane's indirect greenhouse effect, which of the potential methane loss points in the fuel cycles are considered, and the values that have been attributed to these losses. At one extreme (Okken & Kram 1989), a 13% natural gas loss rate is calculated to be needed to offset the CO_2 advantage of natural gas, with actual loss rates put at around 2%. At the other extreme (Abrahamson 1989), natural gas loss rates are estimated to exceed significantly the value calculated to offset the advantage.

A complete analysis of these studies, and the reasons for the disagreement among them, is beyond the scope of this report. The main point to be stressed here is that the uncertainty in the many factors involved is sufficient to allow a very wide range of results, leading to qualitatively, as well as quantitatively, different conclusions being drawn. On the basis of information available regarding the UK situation, however, it seems clear that methane losses from the natural gas fuel cycle are not sufficient to offset its greenhouse advantage over the coal fuel cycle, although better quantification is needed. The study of Eyre (1990) on the gaseous emissions of different electricity fuel cycles would support this conclusion.

It is clear from the foregoing that the question of the relative importance of different greenhouse gas emissions is complex and not fully resolved. Much depends on whether concern is for near or long term changes. It should also be borne in mind that while the longer term effects of a powerful but short-lived gas like methane may be relatively less important, such gases could potentially give rise to more rapid rates of warming in the near term. Relative effectiveness thus depends on whether one is concerned about absolute levels of warming or rates of warming, both of which could have important consequences. As indicated the problem is further compounded when the comparison is taken one step further to focus on the relative effects of different fuel cycles. The attention now being devoted to these questions should considerably clarify this important practical issue in the near future.

REFERENCES

ABRAHAMSON, D. (1989) *Relative greenhouse effect of fossil fuels and the critical contribution of methane*. Paper presented to US Oil Heat Task Force Meeting, 15th June 1989.

BOLIN, B., DOOS, B. R., JAGER, J. & WARRICK, R. A., (1986) *The Greenhouse Effect, Climatic Change and Ecosystems* (Eds) SCOPE 29. J. Wiley and Sons, Chichester.

DAVIES, T. D., KELLY, P. M., BRIMBLECOMBE, P. & GAIR, A. J. (1987) *Surface ozone concentrations and climate: Preliminary analysis*. In proc. WMO Conf. on Air Pollution Modelling and its Applications, Vol 2, WMO/TD No 187, pp 348–356.

DERWENT, R. G., (1990) *Trace gases and their relative contribution to the greenhouse effect*. UK DoE Air Pollution Programme Report, AERE-R13716, Harwell, UK.

EYRE, N. J. (1990) *Gaseous emissions due to electricity fuel* cycles in the United Kingdom. Energy and Environment Paper 1, Energy Technology Support Unit, Harwell Laboratory.

GOLDEMBERG, J., JOHANSSON, T. B., REDDY, A. K. N., & WILLIAMS, R. H., (1985) *An end-use oriented global energy strategy*. Ann. Rev. Energy, 10, 613–88.

KRAUSE, F., BACH, W., & KOOMEY, J., (1989) *Energy Policy in the Greenhouse Vol 1—From warming fate to warming limit: Benchmarks for a global climate convention*. International Project for Sustainable Energy Paths (IPSEP) El Cerrito, California.

LASHOFF, D. A., (1989) *The dynamic greenhouse: Feedback processes that may influence future concentrations of atmospheric trace gases and climatic change*. Climatic Change, 14, 213–242.

LISS, P. S., & CRANE, A. J., (1983) *Man-Made Carbon Dioxide and Climatic Change. A Review of Scientific Problems*. Elsevier.

LOW, P. S., DAVIES, T. D. & KELLY, P. M. (1990) *Uncertainty in surface ozone trend at Hohenpeissenberg*. (Submitted to Nature).

OKKEN, P. A., & KRAM, T., (1989) CH_4/CO_2 *emission from fossil fuels global warming potential*. Paper presented at the IEA/ETSAP—Workshop, Paris, June 1989.

RODHE, H., (1989) *Olika gasers bidrag till växthuseffekten —en jämförelse*. Naturvardsverket Rapport 3647.

SHINE, K. P. (1990) *Effects of CFC substitutes*. Nature, 344, 492–3.

TANS, P. P., FUNG, I. Y. & TAKAHASHI, T. (1990) *Observational constraints on the global atmospheric CO_2 budget*. Science, 247, 1431–8.

WIGLEY, T. M. L. (1988) *Future CFC concentrations under the Montreal Protocol and their greenhouse-effect implications*. Nature, 335, 333–335.

WIGLEY, T. M. L., (1989) In submission by University of East Anglia to House of Lords Select Committee on Science and Technology Inquiry into the Greenhouse Effect. HL Paper 88-II.

WILSON, D., (1989) *Quantifying and comparing fuel-cycle greenhouse gas emissions from coal, oil and natural gas consumption*. Report from the Environment and Energy Systems Studies, Lund University, Sweden.

WORLD COMMISSION ON ENVIRONMENT AND DEVELOPMENT (1987) *Our Common Future (The Brundtland Report)*. United Nations, New York.

Section 3

Emissions of Greenhouse Gases

3.1 GLOBAL EMISSIONS

3.1.1 Carbon Dioxide

Oak Ridge National Laboratory produces annual estimates of world carbon dioxide emissions from fossil fuel consumption and cement production (Marland et al. 1989). Figure 3.1 is based on their estimates for 1950–1986. Total CO_2 emissions more than trebled in that period and now stand at around 5500 M tonnes (expressed as M tonnes of carbon). It can be seen that emissions grew consistently until 1973. The oil price shocks of 1973 and more particularly 1979 caused a decline in emissions. More recently emissions have been rising again and this might partly be related to the fall in oil prices. Within the total, the contribution from oil grew rapidly until 1973 and has since broadly stabilised. The contribution from coal has increased steadily throughout the period whilst that from gas has increased rapidly from a low starting point. Coal and oil each represent around 41 per cent of 1986 total emissions with gas accounting for 15 per cent. The other category in Fig. 3.1 consists of emissions from cement manufacture (2 per cent of the 1986 total) and gas flaring during oil production (1 per cent).

Table 3.1 gives a breakdown of 1986 global CO_2 emissions by country. (The figures are again from Oak Ridge National Laboratory except for the UK where the Oak Ridge figure for that year was found to be in error. An estimate based on the Digest of UK Energy Statistics (Department of Energy 1989) was used instead.) The total emissions and emissions per head are given for all EEC countries and individually for other countries emitting more than 50 M tonnes. The UK total of 160 M tonnes represents 3 per cent of the world total whilst the EEC as a whole represents 13 per cent. The USA is the biggest emitter in the world accounting for 23 per cent of the total, with the USSR accounting for 19 per cent and China a further 10 per cent. On a per capita basis the UK figure is higher than the EEC average of 2.2 tonnes/head of population. The USA figure of 5 tonnes/head lifts the OECD average to 3 tonnes/head. The European Centrally Planned Economies (CPE's) had an average of 3.4 tonnes/head whilst other mainly third world countries average 0·4 tonnes/head.

A comparison of 1986 figures with 1981 (a period of world economic growth) indicates that CO_2 emissions from OECD countries were static, emissions from European CPE's grew at 2 per cent/year and emissions from other countries grew at 5 per cent/year.

An analysis of emissions from electricity generation has been made for this report using International Energy Agency energy consumption data (IEA (1) 1989, IEA (2) 1989). The results indicate that on a global basis 52 per cent of emissions from coal, 9 per cent of emissions from oil and 25 per cent of emissions from gas are from electricity generation. The total emissions from electricity production represent 30 per cent of the global emissions from fossil fuel consumption (and hence around 11–12 per cent of all emissions of greenhouse gases).

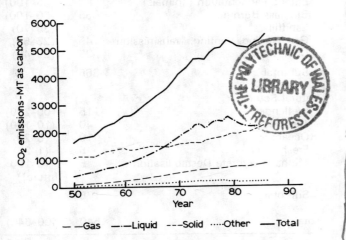

Fig. 3.1. Global carbon emissions 1950–86.

Table 3.1 CO_2 emissions by country (1986) from fossil fuel combustion and cement manufacture

	CO_2 Emissions (Mt C)	CO_2 Emissions (Tonnes C per Capita)
OECD/EEC		
Federal Rep of Germany	186	3.0
United Kingdom	160	2.8
France	98	1.8
Italy	95	1.7
Spain	50	1.3
Netherlands	35	2.4
Belgium	27	2.7
Denmark	17	3.3
Greece	16	1.6
Portugal	8	0.8
Ireland	8	2.2
Luxembourg	2	6.4
Total EEC	702	2.2
Other OECD		
United States of America	1 202	5.0
Japan	256	2.1
Canada	105	4.1
Australia	61	3.9
Other	106	1.2
Total Other OECD	1 730	3.5
Total OECD	2 433	3.0
European CPEs		
USSR	1 011	3.6
Poland	124	3.3
German Democrat Rep	92	5.5
Czechoslovakia	66	4.2
Romania	56	2.4
Other	91	2.0
Total CPEs	1 441	3.4
Other		
China (Mainland)	554	0.5
India	144	0.2
South Africa	93	2.8
Mexico	74	0.9
Brazil	53	0.4
Other	517	0.3
Total Other	1 434	0.4
Total World	5 308	1.1

3.1.2 Methane

Various estimates have been made of the sources of atmospheric methane. Table 3.2 is based on one estimate (Cicerone and Oremland 1988).

Emissions have been divided between anthropogenic and natural sources for the benefit of this report. A small proportion of emissions from enteric fermentation in animals is from wild animals but was not shown separately in the reference used. (Enteric fermentation is essentially the digestion by bacteria of vegetable material such as grass in the stomachs of animals such as cattle and sheep known as ruminants.)

The main sources of anthropogenic emissions are associated with land clearance and food production (principally rice and cattle). Natural gas leakage and coal mining each account for around 10 per cent of anthropogenic emissions. Of the natural sources of methane, wetlands are the major source with termites also making a substantial contribution.

3.1.3 Nitrous Oxide

Table 3.3 gives an estimate of global emissions of nitrous oxide, based on a report by the World Meteorological Organisation (WMO 1987). The WMO report estimated that fossil fuel combustion was a major source of N_2O. However subsequent work by various researchers (e.g. Montgomery et al. 1989, Clayton et al. 1989, Sloan and Laird 1989) has demonstrated that earlier estimates were based on erroneous measurements of N_2O in flue and exhaust gases. Table 3.3 therefore gives a revised estimate for this source substantially lower than the WMO figure (Hill 1990).

The main source of anthropogenic N_2O is estimated to be fertilizer use with biomass combustion

Table 3.2 Estimates of global methane emissions

	M tonnes/year as Methane	
Anthropogenic Sources		
Rice Production	110	(60–170)[a]
Enteric Fermentation (animals)	80	(65–100)
Biomass Burning	55	(50–100)
Landfill	40	(30–70)
Gas, Drilling, Venting, Transmission	45	(25–50)
Coal Mining	35	(25–45)
Subtotal	365	
Natural Sources		
Wetlands	115	(100–200)
Termites	40	(10–100)
Oceans	10	(5–20)
Freshwater	5	(1–25)
Methane Hydrate Destabilisation	5	(0–100 {future})
Subtotal	175	
Total	540	(400–640)

[a] Figures in brackets represent the range of estimates.

Emissions of greenhouse gases

Table 3.3 Estimates of global nitrous oxide emissions

	M tonnes/year as N_2O
Anthropogenic Sources	
Fertilised Agricultural Lands	1·3
Biomass Combustion	1·1
Fossil Fuel Combustion	0·4
Subtotal	2·8
Natural Sources	
Tropical & Sub-Tropical Forests	11·6
Oceans	3·1
Grasslands	2·0
Boreal & Temperate Forests	0·5 (0·2–0·8)
Subtotal	17·2
Total	20·0

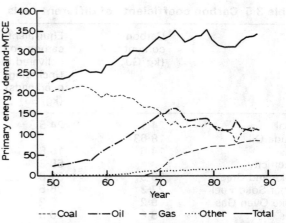

Fig. 3.2. UK primary energy demand 1950–88.

also making a significant contribution. There is some evidence that emissions from vehicles fitted with catalytic converters may be higher than those without so that the contribution from fossil fuel consumption may rise a little in the future.

3.1.4 Chlorofluorocarbons

Estimates of the global production and emission of various CFCs has been provided by one of the main manufacturers, ICI (McCullock 1990). These are shown in Table 3.4.

The first row gives the estimated global production for 1988 whilst the second row gives the cumulative production up to that year. The third row gives an estimate of the cumulative amounts released to the atmosphere whilst the final row gives an estimate of the amounts for CFC11 and CFC12 retained in equipment. These last two pairs of figures were derived by separate accounting methods and only approximately sum to the respective cumulative production.

3.2 UK EMISSIONS OF GREENHOUSE GASES

3.2.1 Carbon Dioxide

Figure 3.2 shows UK primary energy demand from 1950 to 1988 (Department of Energy 1989). It can be seen that total demand rose steadily until 1973 but has broadly stabilised since then. The impact on demand of the oil price rises of 1973 and 1979 and the oil price collapse of 1986 can be clearly seen. Within the total the demand for coal declined until the early 1970's but has since been more stable. The demand for oil grew rapidly until the first oil price shock of 1973 but has since declined. The demand for gas has grown rapidly from the discovery of supplies in the North Sea in the 1960's. The 'other' category comprises nuclear power and hydro electricity. The opposite blips in coal and oil demand in 1984 resulted from the miners' strike of that year.

Table 3.5 shows the carbon coefficients used to convert energy consumption to CO_2 emissions. The first column is the straight forward carbon content of the fuel. The figures for coal and oil are based on the work of Marland and Rotty (1983). The figure

Table 3.4 Estimates of production and emission of various CFCs

Compound	CFC11 (kT)	CFC12 (kT)	CFC113 (kT)	CFC114 (kT)	CFC115 (kT)
Production in 1988	376	421	245	16	16
Total Production to 1988	7 441	9 795	2 191	462	172
Total Emitted to 1988	6 496	9 169	NA	NA	NA
Total Retained in Equipment	1 066	871	NA	NA	NA

Table 3.5 Carbon coefficients of different fuels

	Carbon content fuel (kg/GJ)	Equivalent carbon content delivered fuel (including overheads) (kg/GJ)
Coal	24.1	24.5
Crude Oil	18.63	–
Gas	14.6	15.8
Electricity	–	58.3
Coke and Breeze	28.9	31.8
Other Solid Fuel	27.3	36.5
Coke Oven Gas	9.2	9.2
Gas Oil	18.8	20.5
Burning Oil	18.5	20.1
Fuel Oil	19.8	21.6
Derv	18.8	20.5
Petrol	18.2	19.9
Aviation Fuel	18.5	20.1
Weighted Average (All Oil Products)	18.63	20.3

for gas was produced by British Gas (1990) and represents an average figure for UK North Sea supplies.

The second column of Table 3.5 shows the equivalent carbon content of delivered fuels and is used in subsequent analysis. The figures each contain an overhead for the CO_2 emitted in the production, conversion and (in the case of electricity and gas) delivery of the fuel to the end user. Thus for example the figure for electricity includes the emission of CO_2 at power stations from coal, oil and gas burning and also a proportion of CO_2 from fuel consumed at collieries, oil refineries and by the gas industry.

Figure 3.3 shows UK CO_2 emissions from energy sources. It can be seen that total emissions grew steadily to 1973 but have since declined and reflects the pattern of fuel use throughout the period. The total for 1988 was some 12 per cent below the peak figure of 1973.

Table 3.6 gives a breakdown of UK CO_2 emissions by direct use. The data for emissions from fuel consumption are for 1988. The figure for cement manufacture is for 1986 and is essentially for the CO_2 released in converting limestone (calcium carbonate) into lime (calcium oxide) and excludes the fuel used in the process.

Of the total emissions shown in Table 3.6, 43 per cent are from coal, 37 per cent from oil, 19 per cent from gas and 1 per cent from cement manufacture. The figure for emissions from oil by the oil industry includes 1.35 M tonnes from gas flaring during oil production (included under oil). Thirty-three per cent of total emissions are from power stations and a further 6 per cent are from the other fuel production and conversion industries.

Table 3.6 UK emissions of CO_2 (as carbon) by direct use and fuel type 1988

	CO_2 Emissions (thousand tonnes of carbon)				
	Solid fuel	Oil	Gas	Total	% of total
Fuel producers					
Power Stations	49 069	4 590	128	53 787	33.2%
Oil Industry	0	6 333	0	6 333	3.9%
Gas Industry	0	50	2 220	2 270	1.4%
Coal Mining	121	0	34	154	0.1%
Coke Plants	946	0	3	950	0.6%
Smokeless Fuel Production	235	0	22	256	0.2%
Total	50 370	10 974	2 406	63 750	39.3%
Industry	11 483	8 287	8 338	28 107	17.3%
Transport	0	34 778	0	34 778	21.4%
Domestic	6 176	1 896	15 797	23 868	14.7%
Public Administration	858	2 176	1 913	4 947	3.1%
Agriculture	15	718	46	779	0.5%
Miscellaneous	329	1 119	2 657	4 104	2.5%
Total	18 860	48 973	28 751	96 584	59.6%
Total fuel use				160 334	98.9%
Cement manufacture				1 822	1.1%
Total	69 230	59 947	31 157	162 156	100.0%
Percentage of total emissions	42.7%	37.0%	19.2%	100.0%	

Emissions of greenhouse gases

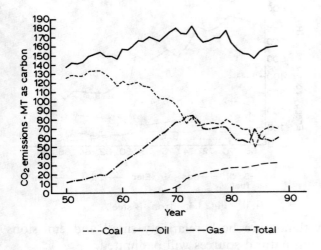

Fig. 3.3. UK CO_2 emissions 1950–88.

Table 3.7 gives a breakdown of CO_2 emissions by end users with all emissions including those for fuel production and conversion, allocated to the end users. Electricity is now shown as a separate fuel and thus emissions of CO_2 at power stations are allocated to the users of electricity. Additionally, emissions of CO_2 from the fuel used (including electricity) in the coal, oil and gas industries are allocated to end users. The figures for emissions from electricity are thus a little different in Table 3.7 compared to Table 3.6. Some electricity is used in the other energy production and conversion industries and appears as part as the overheads for solid fuel, oil and gas. Conversely, the overheads for solid fuel, oil and gas add to the equivalent carbon content of electricity.

Table 3.7 UK emissions of CO_2 (as carbon) by end user and fuel type 1988

	CO_2 Emissions (thousand tonnes of carbon)					
	Solid Fuel	Oil	Gas	Elec	Total	% of Total
Industry						
Iron and steel	6 489	737	1 012	1 713	9 951	6·1%
Non ferrous metal	762	94	259	781	1 896	1·2%
Mineral products	1 170	628	1 053	1 606	4 456	2·7%
Chemicals	1 065	2 609	2 546	3 133	9 352	5·8%
Mechanical engineering	91	494	813	1 881	3 279	2·0%
Electrical engineering	23	232	328	1 076	1 659	1·0%
Vehicles	170	315	496	1 203	2 184	1·3%
Food, drink, tobacco	390	1 126	1 051	1 918	4 485	2·8%
Textiles, leather and clothing	163	295	338	744	1 539	0·9%
Paper, printing and publishing	526	344	544	1 269	2 683	1·7%
Other industries	103	1 857	531	2 266	4 756	2·9%
Construction	0	923	43	224	1 190	0·7%
Unclassified	1 252	0	287	690	2 229	1·4%
Total	12 203	9 035	9 300	18 504	49 043	30·2%
Transport						
Rail	0	644	0	687	1 331	0·8%
Road	0	30 490	0	0	30 490	18·8%
Water	0	975	0	0	975	0·6%
Air	0	5 810	0	0	5 810	3·6%
Total	0	37 919	0	687	38 606	23·8%
Domestic	6 637	2 067	17 057	19 393	45 154	27·8%
Public Administration	887	2 373	2 066	3 968	9 294	5·7%
Agriculture	16	783	50	859	1 707	1·1%
Miscellaneous	360	1 220	2 869	10 263	14 712	9·1%
Export of Oil		853			853	0·5%
Statistical Difference					348	0·2%
Total Fuel Use					160 334	98·9%
Cement Manufacture					1 822	1·1%
Total	20 103	54 014	31 342	53 674	162 156	100·0%
Percentage of Total Emissions	12·4%	33·3%	19·3%	33·1%	100·0%	

It can be seen from Table 3.7 that industry accounts for 30 per cent of total emissions, transport for 24 per cent (of which road transport 19 per cent), the domestic sector 28 per cent and the remaining sectors 18 per cent. The statistical difference shown in Table 3.7 is due partly to statistical differences between tables shown in the Digest of Energy Statistics (Department of Energy 1989) and partly to the allocation of overheads to stock changes of secondary fuels.

3.2.2 Methane

Estimates of UK anthropogenic emissions of methane are published annually by the Department of the Environment (e.g. DoE 1988). Table 3.8 gives the estimates for 1988 which have been provided by Warren Spring Laboratory to DoE (WSL 1990). The figure for emissions from coal mining in the table represents an updated estimate on the WSL figure and has been provided by British Coal (1990).

The figure for landfill in the table is based on the conversion of 25 per cent of the carbon content of the refuse that is tipped into landfill sites into methane. Other bodies have suggested a 75 per cent conversion factor which would indicate a landfill emission figure of around 2 M tonnes.

Other sources also assume higher methane emissions per animal compared to the ones used in the production of Table 3.8 and these would indicate total emissions from animals of around 2 M tonnes. Emissions from cattle and other domestic animals (mainly sheep) are estimated to be the major source of methane emission in the UK. Emissions from the production, distribution and use of energy account for 43 per cent or less of the total.

No estimates are available on emissions of methane from natural sources in the UK. However, the number of wild ruminants and the area of wetlands are both relatively small and emissions from natural sources will be limited.

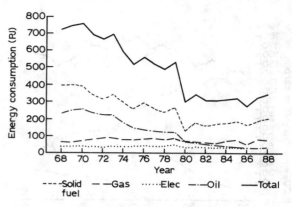

Fig. 3.4.   Iron + steel.

3.3 TRENDS IN ENERGY CONSUMPTION

Future UK CO_2 emissions will depend on the total UK energy consumption, and the breakdown of consumption between fuels. It is not the intention of the Watt Committee here to make forecasts of future energy consumption. These invariably prove to be wrong in the out-turn. This section rather is intended to highlight the trends in energy consumption to indicate what the trends in CO_2 emissions might be. Figure 3.2 which was discussed earlier shows that UK total energy demand grew steadily to 1973, declined following two oil price shocks, and has started to increase since 1985. The pattern however has varied from sector to sector and this is analysed here. Energy consumption is given in this analysis in units of petajoules (PJ) and exajoules (EJ). One PJ is approximately equal to 39 000 tonnes of coal, 22 000 tonnes of oil or 9·5 million therms of gas. One EJ is equal to 1000 PJ.

3.3.1 Iron and Steel

Figure 3.4 shows the trends in energy consumption in the iron and steel industry. Energy consumption

Table 3.8 UK anthropogenic emissions of methane

Source	Emissions MT as methane
Landfill	0·716
Cattle	0·792
Other Animals	0·348
Coal Mines	0·830
Flaring and Venting	0·219
Gas Leakage	0·345
Road Transport	0·021
Other	0·002
Total	3·273

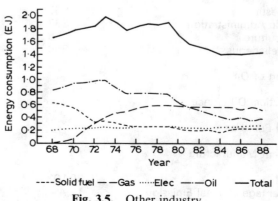

Fig. 3.5.   Other industry.

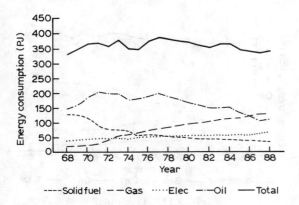

Fig. 3.6. Public administration.

declined up until 1980 as a result of the closure of less efficient plants. Since then total consumption has remained broadly constant with an increase in solid fuel consumption at the expense of oil.

3.3.2 Other Industry

Energy consumption in the other industry sector (Fig. 3.5) reached a peak in 1973, it subsequently declined following the two oil price shocks, but has remained stable in recent years. Over the same period manufacturing output grew steadily to 1973 but declined by 18 per cent up to 1981. Since 1981 there has been strong growth in manufacturing output but this has been accompanied by stable energy consumption.

3.3.3 Public Administration

Energy consumption in the public adminstration sector (Fig. 3.6) reached a peak in 1977. Since then it has declined slowly at an average rate of 1·1 per cent p.a. Over the period gas and electricity consumption have increased at the expense of oil and coal consumption.

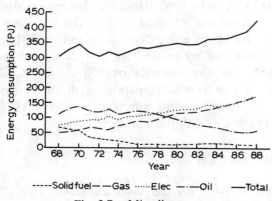

Fig. 3.7. Miscellaneous.

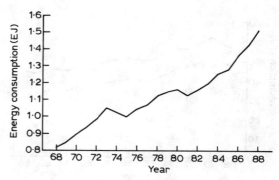

Fig. 3.8. Road transport.

3.3.4 Miscellaneous

The miscellaneous sector includes shops, offices and other premises. Energy consumption in this sector has increased by an average of 1·7 per cent p.a. over the time period shown (Fig. 3.7). During the period gas and electricity consumptions have increased whilst solid fuel and oil consumptions have decreased.

3.3.5 Road Transport

Energy consumption in the road transport sector is almost exclusively from oil. Figure 3.8 shows that energy consumption has significantly increased over the time period shown, with slight decreases at the time of the two oil price shocks. The growth in energy consumption was 3·1 per cent p.a. over the whole period but has been 4·2 per cent p.a. more recently (since 1981).

3.3.6 Air Transport

Figure 3.9 shows energy consumption in the air transport sector, which is met exclusively by oil. Energy consumption has significantly increased over the time period, but the effect of the two oil price shocks is greater than in the road transport sector. Energy consumption has grown at an

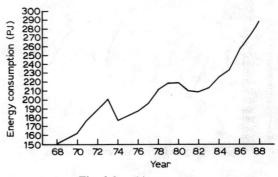

Fig. 3.9. Air transport.

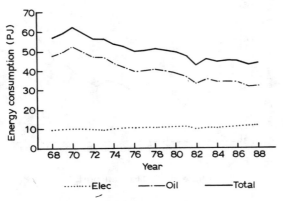

Fig. 3.10. Rail transport.

average of 3·3 per cent p.a. over the whole period, but since 1981 it has grown more rapidly at 4·6 per cent p.a.

3.3.7 Rail Transport

Energy consumption by rail transport has been decreasing since 1970 but has become more stable since 1983 (Fig. 3.10). Energy demand is met by both oil and electricity. Over the time period there has been significant decreases in oil consumption with a small increase in electricity consumption.

3.3.8 Water Transport

Energy consumption in water transport is mainly oil and has remained broadly constant through the time period (Fig. 3.11).

3.3.9 Domestic

The market for domestic energy consumption has been increasing steadily at an average rate of 0·7 per cent p.a., since 1968. This trend is shown in Fig. 3.12. The most significant increase has come from gas consumption with some increase in electricity consumption. Solid fuel and oil consumptions have both declined.

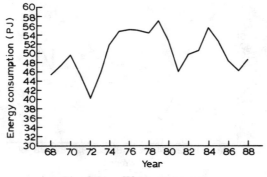

Fig. 3.11. Water transport.

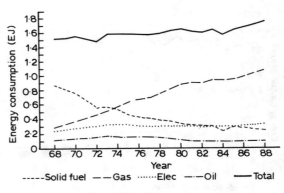

Fig. 3.12. Domestic.

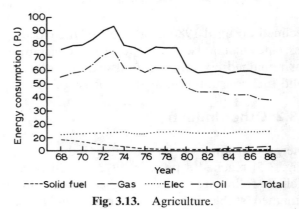

Fig. 3.13. Agriculture.

3.3.10 Agriculture

Agricultural energy consumption declined from a peak in 1973 and has been fairly stable since 1980. Energy consumption is met mainly from oil with electricity providing a significant contribution (Fig. 3.13).

3.3.11 Summary

In summary it is anticipated that energy consumption within the UK will increase and this will lead to increases in CO_2 emissions. The rate of increase in CO_2 is likely to be lower than that for energy due to an increased use of natural gas. On the basis of current trends in energy consumption, emission of CO_2 would rise by around 20% by 2005 compared to 1988. On these trends over half the increase would be due to road transport, with air transport and the domestic and miscellaneous sectors also contributing additional amounts.

REFERENCES

British Gas, (1990). *Personal communication.*
British Coal, (1990). *Personal communication.*

CICERONE, R. J. & OREMLAND, R. S. (1988). *Biogeochemical aspects of atmospheric methane*, Global Biogeochemical Cycles, Vol 2, No 4.

CLAYTON, R., SYKES, A., MACHILEK, R., KREBS, K. & RYAN, J. (1989). N_2O *field study*. Final Report, NTIS PC A05/MF A01.

Department of Energy, (1989). *Digest of United Kingdom Energy Statistics 1989*, HMSO.

Department of the Environment, (1988). *Digest of environmental protection and water statistics*, HMSO.

HILL, T. A. (1990). *Personal communication*.

International Energy Agency, (1989). *Coal Information 1989*, OECD.

International Energy Agency, (1989). *World Energy Statistics and Balances 1971–1987*, OECD.

MARLAND, G. & ROTTY, R. M. (1983). *Carbon dioxide emissions from fossil fuels: A procedure for estimation and results for 1950–1981*, DOE/NBB-0036.

MARLAND, G., BODEN, T. A., GRIFFIN, R. C., HUANG, S. F., KANCIRUK, P. & NELSON, T. R. (1989). *Estimates of CO_2 emissions from fossil fuel burning and cement manufacturing, based on the United Nations energy statistics and the US Bureau of Mines cement manufacturing data*, ORNL/CDIAC-25.

McCULLOCH, A. (1990). *Personal communication*.

MONTGOMERY, T. A., SAMMELSEN, G. S. & MUZIO, L. J. (1989). *Continuous infrared analysis of N_2O in combustion products*, JAPCA 39, No 5.

SLOAN, S. A. & LAIRD, C. K. (1989). *Measurements of nitrous oxide emissions from PF fired power station*, CERL.

Warren Spring Laboratory, (1990). *Personal communication*.

World Meteorological Organisation, (1987). *Atmospheric ozone 1985*. Global Ozone Research and Monitoring Project, Report No 16, WMO.

Section 4

Energy Conversion and the Release of Greenhouse Gases

4.1 INTRODUCTION

If it is accepted that CO_2 emissions from fossil fuel combustion represent a substantial part of the radiative forcing effect, then strategies for lowering these emissions through supply-side technologies will need to be based on a combination of:-

(a) increasing the thermal efficiency of plant
(b) substituting fuels with a lower specific emission rate
(c) increasing the usage of non-CO_2 emitting technologies.

The intention of this chapter is to identify the supply-side energy conversion technologies which are available for limiting the emission of CO_2. This primarily implies discussion of those technologies available for the generation of electricity. No attempt has been made to examine the full systems impact of energy supply or use options, whether these relate to monetary savings, social detriments or greenhouse gas releases. These would include the impacts associated with the materials extraction and fabrication needed to produce plant or equipment, and examined over time i.e. not just instantaneous effects. Only when plant or equipment is replaced at the end of its normal life-span is marginal analysis likely to be reliable. Any holistic analysis would also need to take account of differing ground rules in cost or cost-effectiveness analysis, not only in relation to money values and discount rates but also in terms of contingency allowance, overheads and wider systems cost implications. Load or utilisation factors (and practices) are particularly critical not only for power plants and heat distribution networks but also for domestic equipment.

World wide, fossil fuels produce about 62 per cent of electricity needs with hydroelectric, nuclear and other methods accounting for 19 per cent, 17 per cent and 2 per cent respectively.

4.2 THERMAL EFFICIENCY OF ELECTRICITY GENERATING PLANT

The electricity generating industry has made significant improvements in its specific emission rate (the amount of CO_2 emitted per unit of electricity supplied) over the past few decades. In the years 1950, 1970 and 1987 the Central Electricity Generating Board (CEGB) emitted 18·7, 50·2 and 49·6 Mt of carbon as CO_2 in supplying 48·9, 186 and 228 TWh of electricity respectively. This translates to 0·38 kgC per kWh in 1950, 0·27 in 1970 and 0·22 in 1987. Between 1970 and 1987, the rate of emission of carbon fell by a small amount whilst the number of units of electricity supplied increased by nearly 23 per cent.

The primary reasons for this reduction in specific emissions are the increases in the efficiency of generation plant, changes in fuel mix, the increased use of nuclear power and the use of more sophisticated load management and system control techniques such as at the Dinorwig pumped storage station. Thermal efficiency improvements account for two thirds of the total reduction in specific emissions with the highest thermal efficiency obtained by a coal-fired CEGB unit rising from 30·8 per cent in 1956, to 37·1 per cent in 1980 and to over 38 per cent at present. These improvements have resulted from a variety of factors including increases in the size of plant, improvements in the technologies of control and instrumentation, a change away from stoker towards pulverised fuel firing and improvements in metallurgical technologies which have facilitated the use of higher steam temperatures and pressures. Such changes have been substantial in terms of reducing coal usage and hence CO_2 emissions. For example, for a 2000 MW plant, a yearly reduction in coal consumption of some 155 000 tonnes may be expected per percentage point increase in thermal efficiency.

Whilst present projections assume an improvement in the efficiency of use of fuel, there is a limit to the extent to which this mechanism can be employed to reduce CO_2 emissions. The rate of increase in fuel efficiency which has characterised the industry's performance in the last few decades is unlikely to be maintained in the future without substantial changes in technology. This is because there is a thermodynamic limit to the extent of the improvement which can be obtained. Existing technology, conversion of heat into steam which subsequently drives turbines, has a theoretical maximum thermal efficiency of just over 40 per cent irrespective of the type of fuel or plant used. Small increases beyond the current limit may be possible by using supercritical cycles although the cost-effectiveness of these systems has not yet been conclusively demonstrated. Research is being carried out into those marginal areas where plant efficiency can be increased in a cost-effective manner, typically being directed at changes in the turbines or boilers, low pressure heaters, condensers, cooling towers and burners. The use of atmospheric pressure bubbling or circulating fluidised beds is not likely to generate substantial efficiency gains compared to a new pulverised coal plant since it is constrained by similar thermodynamic considerations.

The energy wasted by conventional power plant operation is rejected to the environment as warm water which has little economic value, although there are industries such as horticulture which have been able to utilise a small amount of this low-grade energy source.

There is believed to be a wide range in the efficiencies of plant operating throughout the world, due to variabilities in local expertise, inabilities to obtain capital for refurbishment and the lack of market incentives to minimise fuel use. There is as a result a high potential for the UK in developing the energy consultancy services required to remedy the situation, particularly through aid programmes or technology transfer to the developing countries. This expertise will become increasingly valuable if the price of fuel rises.

4.3 FUTURE FOSSIL-FIRED TECHNOLOGY

Alternative forms of generation, particularly those such as the combined-cycle, offer the prospect of a reduced specific CO_2 emission rate by using a gas turbine in conjunction with a standard steam turbine. Four possibilities should be considered for combustion of coal each of which could probably be available for commercial exploitation within the next decade:-

— pressurised fluidised bed combustion (pfbc);
— atmospheric pressure circulating bed combustor with an air heater and turbine operating on clean air;
— integrated gasification combined cycle plant based on the British Gas/Lurgi slagging gasifier;
— topping cycle being developed by British Coal involving a combined pfbc/gasification system.

Thermal efficiency data (on a gross CV basis)

Table 4.1 Plant efficiency and CO_2 emissions

Plant Type	Gross Thermal Efficiency (%)	Emissions (kg carbon per GJ generated)[a]
Pulverised Fuel		
Modern 660 MW	38	64·8
Modern 660 MW with fgd	37·5	65·7
Advanced Coal Combustion (Calculated)[b]		
Pressurised fluidised bed	40·4	61·0
Atmospheric circulating fluidised bed	38·2	64·5
integrated gasification combined cycle	40·2	61·3
topping cycle	45·0	54·7
Gas Combined Cycle		
Existing cc gas turbine	42	32·8
Advanced cc gas turbine	48	28·7

[a] Calculated on the basis of coal calorific value of 24 GJ/tonne coal, generating 2·17 tonne CO_2/tonne coal.
[b] Efficiency data from Dept of Energy (1988).

estimated for the UK Department of Energy (1988), are given for these configurations in Table 4.1 from which it may be seen that, of these short term options, only the topping cycle is likely to produce substantial reductions in CO_2 emissions compared with existing pulverised fuel plant. In the longer term, it may be that improvements in gas turbine and steam turbine conditions will raise these efficiency levels by several percentage points.

Magnetohydrodynamics also has the potential for increased efficiency in generation although this technology requires substantial development before becoming economically viable.

In order to make major efficiency increases, it is necessary to treat coal as a chemical feed stock rather than just a fuel: i.e. to look at energy conversion by means other than heat engines. This route could be fuel cells. Studies at the Institute of Gas Technology in Chicago—based on molten carbonate fuel cells, which were the most promising at the time—projected potential efficiencies of over 60%.

4.4 THE SUBSTITUTION OF FUELS

The different chemical compositions of the available fossil fuels imply a different potential for CO_2 emissions. Substitution of natural gas or oil for coal in a cycle of the same efficiency reduces the CO_2 emitted by around 40 and 20 per cent respectively. Existing coal-fired plant may possibly be converted to burn gas, by co-firing or by repowering using an open-cycle gas turbine exhausting into the boiler. It is, however, evident from the data given in Table 4.1 that there would be a substantial benefit, in terms of specific CO_2 emissions, in the use of dedicated gas-fired combined cycle gas turbines (ccgt). Even existing designs of ccgt appear to offer significant efficiency and emission advantages and considerable further improvements are in prospect. The use of natural gas also provide benefits in terms of simultaneous reductions in emissions of sulphur and nitrogen oxides without additional major capital investment.

National Power and PowerGen have announced their intention to commence construction of gas-fired combined cycle plant. Many of the independent generators who have signalled their intention to enter the electricity market on privatisation have also indicated that gas would be their preferred fuel. Although such policies based on fuel switching from coal to natural gas are currently commercially, rather than environmentally, driven they will lower specific CO_2 emission rates in the short term. In the medium to long term, the penetration of such technologies into the generation market might, however, be limited by the extent and geographic location of the available gas reserves, and it is important that adequate research is undertaken into the mechanisms by which supplies of natural gas could be ensured for the future.

4.5 COMBINED HEAT AND POWER (CHP)

Improvements in the CO_2 emission rate may also be achieved by utilising the heat which is currently rejected from the power station in a combined heat and power process. In such a scheme, the energy which is not converted to electricity is fed as hot water or steam to industrial or domestic customers. It has been estimated (Open University 1989) that the provision of coal-fired CHP to 25 per cent of the UK housing stock would result in a reduction of approximately 8 Mt of carbon emissions yearly since overall system energy efficiencies of 70 per cent or more are possible.

CHP has been widely used by large industry for many years with a total installed capacity of around 2 GW (electrical). Many of the installed systems are relatively old with low power/heat ratios. Thus whilst they are responsible for around 40 per cent of industrial steam requirements, the introduction of more modern technology with much higher power/heat ratios would allow substantially higher electricity generation within CHP schemes. This would reduce net emissions of CO_2.

Whilst the potential for increased CHP in industry is large, the reverse appears to be true for district heating. The high capital cost of the infrastructure required to distribute the heat and its relatively low annual average load factor make it difficult for district heating to compete with the existing direct use of, for example, natural gas.

CHP is not however, restricted to large scale use. Systems are being developed and marketed for 15–300 kW of electricity and 40–500 kW of heat, appropriate for small factories. The seasonal and diurnal variations in demand (for both heat and power) do, however, demand a fairly sophisticated management system to optimise energy usage. Further complications are added by the possible requirement to integrate with the national grid supply and to negotiate a realistic price for any exported power.

4.6 CO_2 REMOVAL

In principle, there is a fourth category to add to those delineated in Section 4.1, namely the removal of CO_2 from power plant or industrial flue gases. This has been examined, and indeed has been operated on a small scale, for its potential to separate CO_2 for enhancing oil recovery. Studies of technologies such as selective membrane separation or absorber/stripper systems using an alkanolamine based solvent have been made. CO_2 would be disposed as liquid into the oceans or into other long-term storage locations such as depleted oil or gas wells. The environmental risks of such a process, particularly in ensuring the permanent isolation of the captured CO_2 from the atmosphere would need to be fully covered. The economics of any fossil-fired plant would be seriously affected.

4.7 NUCLEAR ENERGY

The use of nuclear power has the potential to make substantial reductions in CO_2 emissions. Its use in England and Wales over the last few decades has contributed to the improvements in specific emission rate noted in Section 4.2 but major reductions have been achieved in Scotland and France due to the much greater fraction of the generating capacity in these countries being nuclear powered. For example, a 1175 MW PWR such as that being built at Sizewell can be expected to abate about 1.4 Mt of carbon emissions per year relative to the equivalent coal fired capacity.

The cost-effectiveness of this technology is highly dependent on the imposd financing arrangements. Questions will continue to be asked about the costs and risks of nuclear energy but it is possible that perceived benefits will change with technical developments of nuclear systems and increased understanding of the implications of global warming. Nuclear power already exists as a proven, commercial and widely-used means of electricity generation.

It is already making a significant contribution to CO_2 reductions throughout the world. R&D continues towards improving the efficiency of uranium utilisation, into safer designs of reactors, and into the decommissioning and radioactive waste issues that are key to regaining the public's confidence. It is important to maintain nuclear energy as one of the options for future development on a wider scale both in the UK and elsewhere, although it will remain essential that any country wishing to own, operate and eventually decommission nuclear plant must have a sufficient complement of well trained, skilled and experienced staff (see also Chapter 8).

4.8 RENEWABLE ENERGY

On a global scale, the potential for renewable energy is enormous. Table 4.2 indicates that it could, in theory, supply the entire present world electricity consumption of 10 000 TWh/y. However, many of the technologies required to exploit these resources are in their infancy and may prove difficult to introduce through their inherent intermittency, land requirements or environmental impact. There are commentators (e.g. Association for the Conservation of Energy 1989) who have postulated a future global energy scenario of supply from renewable sources feeding high technology and high-efficiency electrical appliances. Such a scenario, if adopted as a goal, would of course have major long-term impacts on fossil fuel industries.

CEGB studies (CEGB 1989) have suggested that the contribution of alternative sources to UK requirements for electricity production might range from about 1 per cent in 2000 to 7 per cent in 2005. Assuming that the various sources become technically and economically viable and publicly acceptable, they might possibly meet 18 per cent of UK electricity requirements in England and Wales by the year 2030. The uncertainties in the studies are large and so these values must be regarded as illustrative only and not as firm predictions. The following paragraphs regarding the details of individual technologies are based on data contained in House of Lords Select Committee on the European Community (1988), Goddard (1988), Chester (1989), and Department of Energy (1989).

4.8.1 Tidal Power

There are many tidal power sites in the UK which are estimated to have the potential to provide about 50 TWh/year of electricity. Amongst the best are the

Table 4.2 World renewable resources

Renewable Source	TWh/year	
	Potential	Realised
Wind	200 000	4
Tidal	200	<1
Wave	4 000	–
Geothermal	>300	35
Hydro	>13 000	2 000
Solar	?	<1

Severn and Mersey Estuaries. According to the recent study by the Severn Tidal Power Group, the Severn could produce 14.4 TWh/year, equivalent to about 6 per cent of current demand in England and Wales. The installed capacity would be 8640 MW, providing a firm power contribution of 1300 MW to the supply system. The Mersey Barrage would be much smaller than the Severn, having a potential output of about 500–600 MW with an annual production of about 1 TWh.

Many other UK estuaries have been considered. Larger schemes include Morecambe (3000 MW), the Solway (7200 MW, approaching the output of the Severn Barrage), the Wash (2400 MW), the Humber (1080 MW) and the Thames (1120 MW). All have been estimated to be of significantly higher cost than either the Mersey or Severn schemes. Smaller estuaries include the Camel at Padstow, and the Loughor and Conwy estuaries in Wales.

4.8.2 Wave Power

The potential wave power resource off the UK coasts is some 50 TWh/year—about 20 per cent of the output of the CEGB in 1989—though the engineering and environmental problems of harnessing even a fraction of this are daunting. No large scale wave energy device has yet emerged which would realistically produce electricity at less than about 10 p/kWh at April 1988 prices, although small scale devices mounted on the shore line, possibly off the north coasts of Devon and Cornwall, may be significantly more cost-effective. The total contribution of power from shore line devices would however be small.

4.8.3 Geothermal Power

In the UK there is no geothermal steam. Geothermal aquifers are a possible source of hot water at temperatures up to about 90°C, but their potential is limited. The hot dry rock process (HDR) is therefore assumed to be the only feasible source of electricity from geothermal sources.

In an HDR system suitable for power generation, cold water would be pumped down one borehole to a depth of about 6 km, and through an artificially created fracture zone where it would heat up to about 180°C. It would then return to the surface through another borehole. The hot water could be used in a thermodynamic power cycle (of overall thermal efficiency of 8–10 per cent because of the low maximum temperature) or for district or process heating.

This technology is at an early stage—only a 2 km deep reservoir has so far been explored. On modest assumptions 10 TWh/year might be exploited in the next century if the technology can be developed economically. The potential sites for geothermal generation in the South West of England include areas such as Bodmin Moor, Dartmoor and Exmoor.

4.8.4 Biofuels and Waste

A potentially important advantage of biofuels is that they can form a beneficial part of the carbon dioxide cycle, provided that their usage is in balance with their production over the short term. On the other hand, biofuels are also capable of producing harmful emissions, and special care must be taken to ensure that this does not happen.

The use of biofuels and waste is somewhat limited in the UK at present. Methane from sewage may be used for heating and at some larger works for electricity production, and waste products are sometimes used as boiler fuel in the timber and paper industries. However, 90% of the UK's domestic and commercial waste is currently dumped as landfill with very little benefit being taken of the energy produced from its decomposition. If this waste were used as a source of energy it is estimated that 9 million tonnes of coal equivalent (Mtce) per year of biofuels may currently be economically exploitable in the UK rising to perhaps 20 Mtce by 2025. Although other processes are possible, only direct combustion or anaerobic digestion are at present regarded as potentially commercially viable. The decomposition of biofuels can produce methane which is itself a significant contribution to global warming. There is thus an additional environmental advantage in burning biomass fuel and extracting the energy rather than allowing methane to escape to atmosphere in, for example, an uncontrolled landfill.

Solid wastes can be converted to a dry form known as refuse derived fuel (RDF) which is easier to handle, transport and burn. Dry wastes which can be incinerated include domestic, commercial and industrial refuse, straw, wood waste and forest thinnings. Five Mtce/year of dry wastes are now estimated to be economically available and the incineration of such material, particularly in a combined heat and power unit, is a potentially commercial option.

Fuels from wet wastes include landfill gas, methane produced from sewage and industrial effluent (including food production), and animal and crop residues. Many applications are economic now. Although landfill for methane production is becoming more widespread, suitable sites are becoming increasingly scarce. Digestion of animal and crop residues on farms is technically practicable, although the economics have so far seemed disappointing. The process converts noxious effluents into useful fertiliser and its use in fully integrated agricultural management may significantly alter the economic picture.

Energy forestry may supply as much as 10 Mtce/year. One option is modified conventional forestry in which trees would grow to maturity over about 25 years. Another option is coppicing, using trees such as willow, aiming for five years to first production from the establishment of a coppice, with harvesting thereafter about every two years. As yet, there are questions concerned with the economics of the supply chain, with disease resistance, particularly of coppicing species, and with soil management. Any substantial exploitation of energy crops would require significant changes in land use—roughly 4000 km^2 of forest or coppice land would be required to supply 1 GW of generating plant producing perhaps 6 TWh of electricity per year.

4.8.5 Hydro Power

Most large scale opportunities for hydro generation in the UK have now been exploited. Present installed capacity in England and Wales is about 120 MW; an additional 50 MW might be installed in England and Wales by the year 2000 with possibly 700–1000 MW in Scotland. This could include low head, small scale, plant of output down to 25 kW or less.

4.8.6 Wind Power

Wind power, which is one of the most promising of the alternative technologies, could provide a technical resource on land of 45 TWh/year. It could be economic now on some windy sites in the UK, and if capital costs continue to decline, wind power may become competitive on less windy sites. Even if capital costs do fall sufficiently, there is some uncertainty concerning lifetime operation and maintenance costs, because large wind turbines have not been in operation longer than four or five years anywhere.

Consultation and planning processes are in hand for the first wind park sites with the expectancy that machines will be erected on the first site in about 1990. Commercial extensions of these or other sites could be a reality by 1995 with up to 1000 MW as wind power being installed by 2005.

The largest potential alternative energy resource in Great Britain is offshore wind power, which could in theory, if not in practice, provide over 200 TWh of electricity per year, compared with the present electricity demand of about 230 TWh/year in England and Wales. Offshore wind turbines could offer environmental advantages over those on land, since they will be less visually obtrusive than machines on land where it is to be expected that large wind farms will raise major local objections on environmental grounds.

Medium sized offshore wind turbines may have most immediate economic potential. A 750 kW prototype is planned for installation 5 km off the North Norfolk coast with an important objective being to establish the true operation and maintenance costs of such offshore wind turbines.

4.8.7 Solar Power

Although developments in solar power will inevitably be implemented initially in countries with sunnier climates than the UK, the potential of this form of power should not be underestimated. In addition to its contribution to the domestic heating load (through passive or active mechanisms), there is long term potential for application in, for example, photovoltaic cells or photocatalysis.

4.9 HYDROGEN

A considerable amount of effort has been devoted to the development of the concept of the hydrogen economy (Bockris & Triner 1980) as a potentially CO_2 free energy source. The hydrogen produced would be transported from the production plant either as a pipeline gas, compressed gas or as liquefied hydrogen. The transport of hydrogen by pipeline presents little difficulty although its low ignition energy and high burning velocity introduce considerable operational hazards. Liquefied hydrogen, in comparison with gaseous hydrogen, offers the advantage of low pressure and low bulk, but containment difficulties arise from the cryogenic nature of the liquefied gas. It can be used as a vehicle fuel both as a compressed gas or as a cryogenic liquid fuel although the volumetric energy density is lower than gasoline. The use of liquefield hydrogen in rocketry has demonstrated that it can be used with reasonable safety in ground installations but its use on a wide-scale basis for transportation is likely to prove difficult.

Whilst utilisation may pose many practical problems, the real stumbling block to a more widespread adoption of the technology has been the hydrogen production stage. The potentially available methods here are:

(i) Electrolysis of water (Scott & Hafele 1989, Worley 1977, Takahashi 1979), the electricity being produced by nuclear, hydro or photovoltaic solar means. Electrolysis also produces oxygen and large scale demands for this product might alter the currently unfavourable economics for this option.

(ii) Thermal dissociation of water by high temperatures in a nuclear reactor or by focused solar energy. In practice a series of chemical reactions is necessary to separate the oxygen and hydrogen and also to provide a chemical carrier which will, at least theoretically, return to start the cycle again without significant losses. Cycles which are being studied constitute basically four families: calcium bromide hydrolysis using mercury (although the use of mercury introduces toxicity considerations); chlorine-steam reaction followed by decomposition of the hydrochloric acid by chromium or vanadium chloride; the iron-chlorine series which has been the subject of extensive analysis; and the most promising process, the thermochemical cycle using the sulphur-bromine and the iodine-sulphur dioxide reactions.

(iii) Production from fossil fuels. One possibility is the HYDRO CARB process based on the two steps of hydrogenation and thermal decomposition. Another is the gasification of coal by steam. These processes which would involve highly complex chemical engineering plant ultimately form carbon dioxide and are not suitable as a climatic control option unless the carbon, containing by-products, is either stored or used for some other synthesis route.

(iv) Photolytic dissociation of water, possibly involving photocatalysis.

4.10 CONCLUSIONS

The above discussion indicates that the main supply side option for reducing greenhouse gas emissions in the near term appears to be the use of fuel switching (i.e. the greater use of gas in combined cycle gas turbines) rather than the development of more efficient coal or oil-fired systems. In the longer term however, there are a range of technologies which could be developed such as many of the renewables which are currently either not economic, or only marginally so, and which will require the provision of market-based incentives or the introduction of stricter environmental regulations before significant market penetration occurs. It is imperative that research, development and demonstration projects are sufficiently well directed and resourced in order to provide a firmer base for their introduction. This is particularly true for the high commercial risk, low emission, technologies which are currently only on the fringes of practicality.

It is also important, from a climate change point of view, that steps are taken to exploit the energy content of domestic and commercial wastes. This may be achieved by incineration or by combustion of methane from landfill sites, it being important that any such methane is not allowed to escape into the atmosphere.

In the long term, questions over the supply of gas may necessitate an increase in the use of coal as a primary energy source. In that case, it would be prudent to ensure the availability of high efficiency, clean coal conversion technologies.

In the future, forms of conversion of sunlight into energy which are now too expensive or seem impracticable may become a reality. For example, the cost of photovoltaic generation may fall sufficiently for it to be widely used, perhaps in the form of an amorphous transparent coating. It may be possible to mimic photosynthesis to produce fuels, or plant breeding and genetic engineering may enable the productivity of energy crops to be significantly increased. Such developments could significantly influence our view of energy supply and demand bearing in mind the value for the annual average density of solar power reaching the UK is about $100\,MW\,km^{-2}$ (Clapham, Pers. Comm.).

The hydrogen economy was proposed as a method of providing a substitute for oil and gas when they become exhausted. In principle it would also be of assistance in reducing carbon dioxide emissions. At the present time neither nuclear energy nor solar energy are capable of acting as the large-scale sources of electricity or high-temperature heat required to produce the hydrogen from water. In the short term hydrogen could be produced from hydroelectric sources, in the medium term by electrolysis using solar energy and in the long term by solar photolysis of water using solid catalysts (photocatalysis). The House of Commons Select Committee on Energy (1989) has suggested that further research is required in this area. The pressures of global warming may accelerate the pace of change here and it may be worth examining the engineering aspects of basic systems in order to establish the technological problems such as safety, reliability, etc.

If it is accepted that reductions in CO_2 emissions are required, then this implies displacement of large quantities of capacity installed for electricity generation. The time required to implement such a change should not be under-estimated since all generating systems need to undergo a process of engineering studies and tendering, environmental impact analysis, land acquisitional, and planning consent negotiation before construction can start. Experience over the last few years indicates that substantial periods of time will be necessary to bring to fruition many of the technologies outlined above.

It is also worth while briefly mentioning the wide range of unproven measures which have been proposed to provide global energy needs without associated CO_2 production. These range from the very costly to the extremely speculative. Typical examples include fusion energy and satellites in orbit collecting solar energy and transmitting to the ground via electronic beam. It has also been suggested (Siefritz 1989) that the global disruption of climate may be mitigated by placing mirrors in orbit to reflect some of the incoming radiation.

REFERENCES

Association for the Conservation of Energy, (1989). In 6th Report of the House of Commons Energy Committee, HMSO.
BOCKRIS, J. O. M. & TRINER, A. (1980). *Energy Options*, Taylor and Francis Ltd, London.
Central Electricity Generating Board, (1989). In 6th Report of the House of Commons Energy Committee, HMSO.
CHESTER, P. F., (1989). *Prospects for alternative energy sources in electricity generation*. Power Engineering Journal of the EEC. Vol. 3, No. 4, 175–84.
CHESTER, P. F. (1989). *Global Warming in the Worldwide Challenge to Electric Utilities*. Paper given to the US Energy Association Global Climate Change Focus, Washington, 21 November.
CLAPHAM, V. M. (Pers. Comm.). Private contribution to the Working Party 1990.
Department of Energy, (1988). *Prospects for the use of advanced coal based power generation plant in the United Kingdom*, Energy Paper No. 56, HMSO.
Department of Energy, (1989). *An evaluation of energy related greenhouse gas emissions and measures to ameliorate them, Intergovernment Panel on Climate Change*, Department of Energy Paper No. 58.
GODDARD, S. E., (1988). Programme of Evidence on Comparison of Non-Fossil Options to Hinkley Point 'C'. Hinkley Point 'C' Power Station Public Inquiry, CEGB, paragraph 100.
House of Commons Select Committee on Energy, (1989). *Energy policy implications of the greenhouse effect*, Sixth Report, Session 1988–89, HMSO.
House of Lords Select Committee on the European Community, (1988). *Prospects for alternative energy sources in electricity generation*. In Session 1987–88, 16th Report with Evidence. HL Paper 88, pp 17–24.
Open University, (1989). In 6th Report of the House of Commons Energy Committee, HMSO.
SCOTT, D. S. & HAFELE, W. (1989) *The coming hydrogen age: Preventing world climate disruption*, 14th Congress of the World Energy Conference.
SEIFRITZ, W. (1989) *Mirrors to halt global warming?*, Nature, 340, 603.
STEINBERG, M. & CHENG, H. C. AAAS Annual Meeting, Boston, MA.
TAKAHASHI, T. (1979). *Solar Hydrogen Energy System*, Ed. T. Ohta, Pergamon Press, New York.
WORLEY, N. G. (1977). *A review of the potential use of nuclear reactors for hydrogen production*. Gas Engineering and Management, October 371.

Section 5

Energy Usage in the Home, Commerce and Industry, and its Effect on the Release of Greenhouse Gases

5.1 DOMESTIC SECTOR

5.1.1 Introduction

The domestic sector accounted for 28 per cent of all energy delivered to UK consumers in 1988. This proportion is similar to that for 1960 and has shown little variation in the intervening years, having fallen to 25 per cent in the early 1970s and risen again after 1980. This apparent stability is deceptive, however, since considerable structural changes have taken place during that time.

Figure 5.1 shows the amounts of various fuels delivered to the domestic sector from 1970 to 1988 and illustrates the rapid penetration of natural gas, replacing town gas but also displacing solid fuel. Natural gas now dominates the domestic heating market with about 70 per cent of households using it for their main form of heating. Electricity consumption has remained fairly constant throughout the period, a steady growth in usage for lighting and appliances being offset by a decline in usage for heating. Oil has never held a significant market share in the UK domestic sector and has declined since the mid 1970s to its present level of about 6 per cent of delivered energy.

There has also been a very large increase in the use of central heating. In 1970, only 35 per cent of houses had central heating, growing to 70 per cent by 1988. Surveys and field trials have shown that centrally heated houses are significantly warmer than others, on average, and there is no doubt that many houses are now much better heated than they were in 1970. Insulation standards have also improved in new buildings by successive revisions to the Building Regulations while many existing buildings have had insulation added.

The domestic sector was the largest consumer of natural gas in 1988, accounting for 56 per cent of total UK deliveries (cf 48 per cent in 1978). It is also a very significant consumer of electricity (36 per cent in 1988).

Although domestic use of solid fuels is declining, it still accounted for 34 per cent of solid fuels delivered to final consumers (i.e. excluding deliveries to the industry and for electricity generation). The domestic market for petroleum is only 4 per cent of the total UK oil market.

5.1.2 End Uses of Energy Within the Domestic Sector

The following analysis draws largely on the work of the Building Research Establishment (Shorrock & Henderson 1990). Figure 5.2 shows a breakdown of domestic energy consumption by end-use. Space heating is estimated to account for the largest proportion (61 per cent), followed by water heating (22 per cent). Cooking and electricity consumed by lights and appliances account for 7 per cent and 10

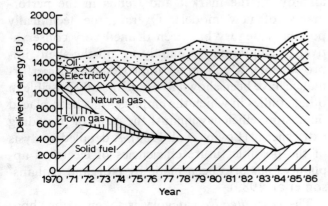

Fig. 5.1. Breakdown by fuel.

per cent respectively. Space heating is predominantly served by natural gas with lesser contributions from solid fuels, oil and electricity. Water heating follows a similar pattern, often being supplied from the same boiler as space heating. Cooking is split between gas and electricity with only small contributions from other fuels. Lights and appliances are of course served almost exclusively by electricity. This last category is specially significant because it has shown a steady growth throughout the last two decades, contrasting with the other end-uses which have remained relatively stable. It also accounts for a relatively larger proportion of energy expenditure than of quantity, reflecting the high primary energy ratio for electricity and the high cost of peak rate electricity.

5.1.3 Carbon Dioxide Attributable to the Domestic Sector

CO_2 is emitted when fossil fuels are consumed by heating appliances within dwellings and also when the electricity used in dwellings is generated. Both types of emission need to be included in any analysis of emissions that is used to explore the impact of improvements in the efficiency of end-use. For the purposes of this study, it has been assumed that the CO_2 emissions associated with electricity use in the domestic sector are the same as the average for all electricity distributed by the national grid. Figure 5.3 shows CO_2 emissions for the domestic sector, broken down by end-use and fuel.

Compared to the energy breakdown in Fig. 5.2, Fig. 5.3 shows that end-uses which rely on electricity are much more significant in terms of CO_2. In fact, electricity for lights and appliances becomes about as significant as all the natural gas used for space heating. However, space heating remains the largest single contributor to CO_2 emissions.

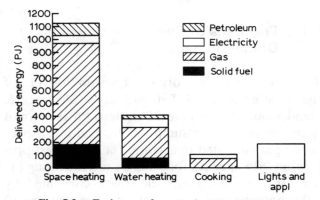

Fig. 5.2. End uses of energy in UK dwellings.

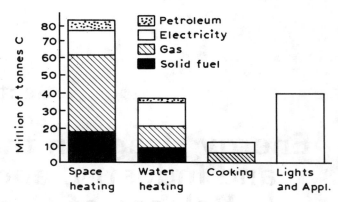

Fig. 5.3. Carbon dioxide emission attributable to UK dwellings, by end use and by delivered fuel.

5.1.4 Energy Efficiency Increases

5.1.4.1 Cost-effectiveness

The following analysis deals only with measures which are technically well proven, but distinguishes between measures which are presently cost-effective and those which are not. There are many possible criteria for deciding what to include in the cost-effective category and it is not intended to define those criteria rigorously. The aim, rather, is to give an indication of the reductions in CO_2 emissions which could be achieved using measures which are generally accepted as cost-effective and are already available for routine application. For this purpose, the following broad definitions have been adopted.

The term *technically possible* is taken to mean what could be done using standard techniques and readily available materials. It does not, however, include retrofitting of measures that would cause excessive disruption, e.g. ground floor insulation in existing occupied dwellings. For household appliances, what is technically possible is difficult to determine and inextricably linked to market conditions. Judgements about what should be included are therefore tempered by consideration of what is already on the market and trends in the performance of new models. Overall, the technically possible category has been defined very conservatively, being confined mainly to 'off-the-shelf' improvements. Much larger improvements in performance have been shown to be technically possible in other studies in which the emphasis has been placed on costing efficiency against the costs of providing extra energy (Lovins et al. 1989). A detailed analysis of the possible future efficiencies of electrical appliances has also been made by Norgand (Johansson et al. 1989).

The *cost effective* category is taken to be those measures from the technically possible category which are currently considered to be a good investment. There are clear difficulties in deciding what to include since individual cases are dependent on existing levels of insulation, current and future fuel costs and opportunities arising through other work being carried out on the building. In the case of double glazing, it is particularly complicated since energy saving is only one of the benefits deriving from its installation. In view of the strong popularity of double glazing as a general home improvement, and the low marginal cost of double over single glazing when windows are being replaced, it has been included in the cost-effective category.

For household appliances what is cost-effective is often the same as is technically possible under our definition. This is because the figures are based on what is available on the market today, selling in direct competition with less efficient equipment and at a similar price. If the production of the more efficient models were significantly more costly than others, they would be at a serious competitive disadvantage: this implies that it is already economic to produce the more efficient models. For other types of appliance the cost-effective savings might be expected to be smaller than could derive from our definition of technically possible. Compact fluorescent lighting, for example, is not likely to be cost-effective for applications in which there is very intermittent usage of low wattage bulbs. There may also be performance considerations which limit uptake, for example the weight and appearance of compact fluorescent lighting. Table 5.1 lists the assumptions used to quantify technically possible and cost-effective measures.

5.1.4.2 Reductions through insulation and draught proofing

Table 5.2a shows the reductions in energy use that could result from applying the technically possible measures for improving thermal insulation and draught proofing as specified in Table 5.1. Table 5.2b shows the equivalent figures for the cost-effective measures. The reductions resulting from insulation measures in dwellings depend upon a number of factors, particularly the standard to which the dwelling is heated. Improvements in standards of heating usually occur whenever insulation is improved, particularly if the standard is poor before the insulation is installed. As standards of heating improve, however, there is a tendency for more of the benefits to be realised in the form of reduced energy usage.

The biggest potential savings from insulation are for walls. Table 5.2a considers savings for all wall types, whereas Table 5.2b considers only those cavity walls which are suitable for cavity filling,

Table 5.1 Assumptions concerning technically- and cost-effective savings. The assumptions used regarding the quantification of technically possible and cost-effective improvements are summarised below for all the energy efficiency measures considered

Energy Efficiency Improvement	Technically Possible	Cost Effective
Insulation improvements	All walls insulated All lofts insulated to 150 mm	80% of all cavity walls insulated Lofts with ⩽25 mm insulated to 150 mm
	Full double glazing in all homes Draught proofing all homes	Full double glazing in all homes Draught proofing in homes with <80% rooms already treated
	80 mm insulation to all hot water tanks	All tanks with <50 mm insulated to 80 mm
Heating efficiency improvements	Condensing boilers in all gas centrally heated homes	Condensing boilers in all gas centrally heated homes
Improvements to cookers	25% efficiency improvement to gas and electric cookers	13% efficiency improvement to gas and electric cookers
Improvements to lights and appliances	20% efficiency improvement to dishwashers 75% reduction in lighting consumption 75% efficiency improvement to refrigeration equipment 20% efficiency improvement to washing machines and tumble driers 25% efficiency improvement to televisions	10% efficiency improvement to dishwashers 38% reduction in lighting consumption 50% efficiency improvement to refrigeration 20% efficiency improvement to washing machines 25% efficiency improvements to televisions

Table 5.2a Technically possible energy savings through application of insulation and draught proofing measures to United Kingdom dwellings

Fuel	Fuel Savings (PJ)					
	Loft Insulation	Wall Insulation	Double Glazing	Draught Proofing	Water Tank	Ideal Saving
Solid	8·6	54·0	16·3	10·2	8·7	97·8
Gas	37·3	235·3	71·1	44·5	22·7	410·9
Electricity	2·8	17·9	5·4	3·4	5·7	35·2
Petroleum	3·6	22·9	6·9	4·3	2·5	40·2
Total	52·3	330·1	99·7	62·4	39·6	584·1

Table 5.2b Cost effective energy savings through application of insulation and draught proofing measures to United Kingdom dwellings

Fuel	Fuel Savings (PJ)					
	Loft Insulation	Wall Insulation	Double Glazing	Draught Proofing	Water Tank	Ideal Saving
Solid	5·9	26·7	16·3	10·1	5·8	64·8
Gas	25·6	116·2	71·1	44·2	15·1	272·2
Electricity	1·9	8·8	5·4	3·4	3·8	23·3
Petroleum	2·5	11·3	6·9	4·3	1·7	26·7
Total	35·9	163·0	99·7	62·0	26·4	387·0

estimated to be about 80 per cent of existing cavity walls. The insulation of solid walls is not taken to be generally cost-effective. It can, however, be cost-effective when it accompanies major refurbishment work. Reductions in emissions deriving from other insulation measures are lower, either because there is little remaining potential for applying those measures or because the measures themselves have a smaller effect.

Tables 5.3a and 5.3b show the reductions in CO_2 emission which would result from the energy reductions shown in Tables 5.2a and 5.2b, assuming no

Table 5.3a Potential reduction in carbon dioxide emissions through the application of technically possible insulation and draught measures to United Kingdom dwellings

Fuel	Reduction in CO_2 Emissions (million tonnes C)					
	Loft Insulation	Wall Insulation	Double Glazing	Draught Proofing	Water Tank	Total Saving
Solid	0·22	1·40	0·42	0·27	0·23	2·54
Gas	0·59	3·70	1·12	0·70	0·36	6·47
Electricity	0·16	1·03	0·31	0·20	0·33	2·03
Petroleum	0·07	0·46	0·14	0·09	0·05	0·81
Total	1·04	6·60	1·99	1·25	0·96	11·05

Table 5.3b Potential reduction in carbon dioxide emissions through the application of cost effective insulation and draught proofing to United Kingdom dwellings

Fuel	Reduction in CO_2 Emissions (million tonnes C)					
	Loft Insulation	Wall Insulation	Double Glazing	Draught Proofing	Water Tank	Total Saving
Gas	0·40	1·83	1·12	0·70	0·24	4·28
Electricity	0·11	0·51	0·31	0·20	0·22	1·35
Petroleum	0·05	0·23	0·14	0·09	0·03	0·54
Total	0·71	3·26	1·99	1·25	0·64	7·85

change to the present mix of fuels consumed. Insulation measures in the housing stock are seen to be capable of reducing UK CO_2 emissions by as much as 12 million tonnes (as C), or 7 per cent of total UK emissions. Cost-effective measures account for about 7·9 million tonnes (5 per cent of total UK emissions).

5.1.4.3 Reductions due to improvements in heating appliance efficiencies

Heating appliance efficiencies have improved over the years for a number of reasons. Firstly, more modern appliances are generally more efficient than their predecessors. Secondly there has been a shift away from open coal fires towards gas heating. The third reason is the growth in the use of central heating, which apart from providing a means of heating the whole dwelling, usually has a higher efficiency than individual heaters. Further improvements in efficiency are likely to occur as existing boilers and other heating appliances are replaced.

A particular development in the design of gas central heating boilers offers considerable scope for improved efficiencies. In a condensing boiler heat is recovered from the flue gases, including that from the latent heat of condensation. This allows them to operate at a much higher efficiency than conventional boilers which are designed to avoid condensation by maintaining a high temperature in the flue. Field trials have shown that condensing boilers can achieve an annual average efficiency of 85 to 90 per cent, compared with 65 to 70 per cent for a conventional boiler. Old and poorly controlled conventional boilers are likely to have very low efficiencies, often in the range of 50 to 60 per cent. Thus substantial reductions in energy use can be made through boiler replacement.

Improved control equipment can optimise the use of energy by providing better time and temperature control in rooms and by preventing excessive boiler cycling. Savings of 5 to 10 per cent are feasible.

The reductions in energy use that can be achieved through heating appliance efficiency clearly depend on how much demand there is for heating which, in turn, depends upon how well the housing stock is insulated. Thus the savings from boiler efficiency improvements would be greater if the insulation was poorer and vice versa. The result is the apparently paradoxical result that for our cost-effective case, the reduction in energy use from boiler efficiency is greater than for the technically possible case. In reality, of course, insulation and boiler replacement could occur in either order and it is not important to assign reductions uniquely to each measure, only to ensure that the overall result is fairly assessed.

The results show that replacement of present boilers with condensing types could reduce CO_2 emissions by 2·3 million tonnes (as C) if no change were made to present insulation standards. If the cost-effective insulation measures were applied, this would reduce to 1·5 million tonnes and to 1·1 million tonnes for the technically possible case. Further reductions are likely to arise from improvements to the efficiency of other heating appliances but will almost certainly be small compared to the reductions possible through the use of condensing boilers.

5.1.4.4 Reductions due to improvements in lights, appliances and cooking

The importance of domestic equipment which consumes electricity has been noted above, with lights and appliance use estimated to account for about a quarter of all domestic related CO_2 emissions. Table 5.4 shows the most important energy consuming appliances, together with estimates of possible energy and CO_2 reductions from applying the efficiency improvements set out in Table 5.1. Refrigeration and home laundry equipment are heavy energy users and so offer the greatest scope for reductions. Lighting is also very important, since apart from being a significant energy user, large gains in efficiency may be obtained by substituting compact fluorescent bulbs for incandescent filament bulbs.

Total reductions in CO_2 emissions due to lighting and appliance efficiency improvements are estimated at 3·0 million tonnes for the technically possible case and 2·2 million tonnes (as C) for the cost-effective case. A further reduction of 0·8 million tonnes could be obtained from improvements in the efficiency of cookers, half of which would fall into the cost-effective category.

5.1.5 Overall Reductions in CO_2 Emissions from Dwellings

Energy requirements in the domestic sector can therefore be reduced by applying a variety of energy efficiency improvements. Space heating needs can be reduced by insulation of the external fabric of the building or by reducing ventilation heat losses. Water heating energy requirements can be reduced by insulation of storage tanks. Both space and water heating energy needs can be reduced by in-

Table 5.4a Energy use in dwellings for lights and appliances and possible savings through efficiency improvements (technically possible case)

Appliances	Typical Appliance Consumption (KWh/year)	Owner Level	National Energy Consumption (PJ)	Technically Possible Efficiency Improvement	Energy Saving (PJ)	CO$_2$ Reduction (million tonnes C)
Washing Machines	200	86%	14.1	20%	2.8	0.16
Tumble Driers	300	31%	7.5	20%	1.5	0.09
Dishwashers	500	7%	2.9	20%	0.6	0.03
Refrigerators	300	57%	14.3	25%	3.6	0.21
Fridgefreezers	740	43%	25.9	25%	6.5	0.37
Freezers	740	39%	23.2	25%	5.8	0.33
Kettle	250	86%	17.7			
Irons	75	98%	6.0			
Vacuum Cleaners	25	98%	2.0			
Televisions	235	98%	19.0	25%	4.8	0.27
Lights	360	100%	29.8	75%	22.4	1.29
Miscellaneous	240	100%	19.8			
Total			182.3		47.8	2.76

Table 5.4b Energy use in dwellings for lights and appliances and possible savings through efficiency improvements (cost effective case)

Appliances	Typical Appliance Consumption (KW/year)	Owner Level	National Energy Consumption (PJ)	Cost Effective Efficiency Improvement	Energy Saving (PJ)	CO$_2$ Reduction (million tonnes C)
Washing Machines	200	86%	14.1	20%	2.8	0.16
Tumble Driers	300	31%	7.5	0%		
Dishwashers	500	7%	2.9	10%	0.3	0.02
Refrigerators	300	57%	14.3	25%	3.6	0.21
Fridgefreezers	740	43%	25.9	25%	6.5	0.37
Freezers	740	39%	23.2	25%	5.8	0.33
Kettle	250	86%	17.7			
Irons	75	98%	6.0			
Vaccum Cleaners	25	98%	2.0			
Televisions	235	98%	19.0	25%	4.8	0.27
Lights	360	100%	29.8	38%	11.3	0.65
Miscellaneous	210	100%	19.8			
Total			182.3		35.0	2.02

stalling heating appliances, such as boilers, which have higher efficiencies. Lights, cookers and electrical appliances can all be replaced with more efficient types. There are, in addition, a number of other means of saving energy which can be termed 'good housekeeping', for example turning off services when they are not required and choosing appliances which are appropriately sized for the needs of the household. These good housekeeping measures are difficult to quantify, being dependent on the lifestyle of individual households. They are nevertheless important since they are related to both public awareness of the significance of energy use and the perceived cost of the energy to the household.

Taken together, all of the potential reductions in CO$_2$ emissions for the domestic sector amount to 12 million tonnes for the cost-effective case and 16.9 million (as C) for the technically possible case. These amounts are equivalent to 7 per cent and 10 per cent of total UK emissions at present. About two thirds of the possible reductions are associated with improvements to thermal insulation and the remaining third with improvements in the efficiency of heating systems, electrical appliances and lighting.

5.1.6 Other Ways of Generating Heat and Power

All of the above reductions assume no change to the mix of fuels used, particularly in the case of electricity generation. This is considered to be entirely appropriate in the context of assessing the potential effects of measures to improve the efficiency of energy use in the domestic sector. There are some possibilities for fuel switching within the sector, for example using gas instead of electricity or solid fuel for heating. The scope for CO_2 reductions by this means is rather small as only a small part of the heating load is met by electricity and much of that could not be readily switched to gas.

A further possibility is the use of combined heat and power in which the heat from the generation process would be distributed for domestic heating. Although this form of generation is widely used in other countries (notably Denmark), the prospect for its introduction into the UK is very limited in the short term. Feasibility studies of the long term potential for large scale CHP suggest that it is capable of achieving net national energy savings of 130 PJ, equivalent to 2·7 million tonnes of CO_2 (as C).

5.1.7 Possible Future Emissions from the Domestic Sector

The foregoing estimates of potential reductions have all been made relative to today's demand and pattern of consumption. A realistic projection of future emissions would also have to take account of likely changes in demand. The number of households is expected to grow and with it the total floor area of dwellings to be heated. The ownership of various energy consuming appliances is also expected to grow: some appliances such as televisions may be close to saturation while others such as dishwashers have a low market penetration in the UK compared to other, similar countries. The ownership of central heating continues to rise, leading to better standards of space heating. Set against those trends, the efficiency of equipment is generally improving with time as is the level of thermal insulation in the housing stock through the building of new, better insulated homes and the refurbishment of existing ones.

5.2 COMMERCIAL AND PUBLIC BUILDINGS SECTOR

5.2.1 Introduction

Energy usage in the commercial and public buildings sector (CPBS) in the year 1985 was studied in detail by the Energy and Technology Support Unit (ETSU) and reported in Herring et al. (1988). The system of building and activity classification used in that report has been maintained but the total energy usage has been updated to 1988 and the associated emissions of CO_2 calculated.

In the CPBS there are a large number of premises of widely varying sizes (see Table 5.5), in which 60 per cent of the energy is consumed for space heating, followed by lighting at 10 per cent, water heating and cooking about 8 per cent each and the remaining 14 per cent for other uses as shown in Fig. 5.4 (Chisholm 1989). A more detailed breakdown of energy consumption by fuel type and end use is given in Tables 5.6 and 5.7 respectively.

The CPBS sector includes buildings and activities within them but excludes manufacturing or transport. Taking the commercial sector first, we find that in 1988 electricity and gas each provided about 40 per cent of the energy, oil and coal the remaining 20 per cent. In the public sector, gas provided nearly 38 per cent of the energy used, followed by oil at 33 per cent with electricity and solid fuel the remainder (DoEn 1989b). The trend in both sectors is towards increasing consumption of electricity and gas and a reduction in oil and coal.

Table 5.5 Estimated number and area of premises within the major commercial market sectors in 1985

	Number of Premises (000)	%	Area (km²)	%
Offices, Insurance and Banking	220	14	54	9
Retail	745	48	208	30
Health	29	2	43	6
Hotels	43	3	38	6
Catering	150	10	30	4
Education	42	3	105	15
National, Local Government, Defence	53	3	101	15
Other	270	17	98	15
Total	1 552	100	688	100

Source: Dept of Energy (1989b).

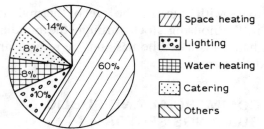

Fig. 5.4. Commercial market energy profile (1985).

5.2.2 The Commercial Section

Offices: This sector grew rapidly by some 19 per cent between 1975 and 1985. Over half are in converted residential buildings more than 40 years old. Offices built in the 1960's were highly glazed and lightweight constructions, with little regard for energy conservation. Over 90 per cent of offices are

Table 5.6 Estimated energy consumption by fuel type in 1985 (units: PJ)

	Electricity	Gas	Oil	Solid	Total	Area (km^2)	SEC (GJ/m^2)
Office	27	21	6	4	58	65	0·90
Distribution	16	50	23	1	90	124	0·73
Shops	42	17	7	3	69	84	0·83
Catering	10	9	a	a	19	9	2·16
Pubs/Clubs	15	17	4	5	41	21	1·96
Residential	18	22	10	4	55	38	1·43
Retail Services	4	2	1	a	8	15	0·51
Other	21	16	8	1	47	84	0·55
Sub Total	152	156	60	19	387	440	0·88
National Government	12	14	9	5	41	33	1·24
Defence	5	7	18	3	33	16	2·06
Local Govt.	11	40	21	5	76	52	1·46
Education	13	34	56	8	111	106	1·05
Health	13	29	40	21	104	43	2·41
Street Lighting	8	–	–	–	8	–	–
Sub Total	62	125	144	43	373	249	1·50
Service	214	281	204	62	760	689	1·10

a Less than 0·5 PJ.
Source: ETSU estimate.

Table 5.7 Estimated energy consumption by end use in 1985 (units: PJ)

	Space Heating	Water Heating	Light	Cooking	Other	Total
Office	35	2	9	1	11	58
Distribution	76	3	6	a	5	90
Shops	37	1	12	1	19	69
Catering	5	2	2	11	a	19
Pubs/Clubs	22	5	6	3	6	41
Residential	25	9	6	11	5	54
Retail Services	5	a	2	a	a	8
Other	28	3	6	1	9	47
Sub Total	233	24	49	27	55	387
National Government	28	2	3	2	7	41
Defence	21	6	3	2	1	33
Local Govt.	58	7	5	3	3	76
Education	85	13	6	6	2	111
Health	59	27	4	8	5	104
Street Lighting	–	–	8	–	–	8
Sub Total	251	55	28	22	18	373
Service	484	78	76	48	73	760

a Less than 0·5 PJ.
Source: ETSU estimate.

rented with the result that the costs of introducing energy efficiency measures may be difficult to recover.

Offices built since the oil crises of the 1970's are better insulated. However, while thermally more efficient, the energy consumption of newer or refurbished offices is often higher because of the prevalence of air conditioning. The use of air conditioners is set to increase as working conditions improve and the recent warm summers may accelerate that trend. In addition, the modern office contains an extensive suite of electrically operated equipment, including computers and their peripherals, typewriters, photocopiers etc.

Improvements in energy efficiency for offices are possible through the use of better insulation and heating systems. Of increasing importance is the minimisation of air conditioning load through building design and selective operation whilst at the same time avoiding the 'sick building' syndrome. The efficiency of electrical appliances is important too since they not only consume power but also contribute to the air conditioning load.

Distribution: Space heating constitutes the main energy usage in warehouses and wholesale properties but air conditioning and refrigeration are increasingly employed. The latter trend has led to increased energy consumption. There is scope for considerable improvements in energy efficiency through the better insulation of such buildings.

Shops: The major fuel used is electricity, the trend towards foods which require extensive refrigeration and chilled display units have increased consumption. New and refurbished stores have higher lighting levels and air conditioning is becoming more common. As a result, while more efficient use is made of energy for space and water heating, energy consumption has not fallen because of increased demand for other services.

Catering: Cooking is the prime energy use and fast food outlets in particular are very energy intensive. There is a large scope for improved energy efficiency, particularly in heat recovery and better control of cooking equipment. However, the main priorities of the catering industries are speed and convenience and until energy efficient equipment meets such needs only a small part of the potential savings are likely to be realised.

Public Houses and Clubs: The numbers of public houses decreased by 10 per cent between 1977 and 1985 but is now stable, as is the number of clubs. However, the average size has increased since it was the smaller establishments which closed. Few new public houses are likely to be built in the next 10 years but refurbishment of many is likely. There is significant potential for energy saving by efficiency improvements in space and water heating. Major breweries are looking to reduce energy consumption in public houses.

Residential Accommodation: Hotels, boarding houses, holiday accommodation, conference centres and residential homes are included here. Hotels account for most of the energy demand and there is the opportunity for introducing many energy saving measures, particularly during refurbishment. However, total energy consumption can increase if air conditioning is installed.

New CHP (combined heat and power) units are being installed in hotels where there is a reasonably constant need for power, hot water and heating. The units are proving economic to run and offer substantial energy savings. With maximum utilisation of the heat produced by a gas-fired CHP unit, only one third of the energy is used and one quarter of the CO_2 emitted compared with electricity and heating supplied by conventional means.

Retail Services: A wide area of activities are covered in this sector such as vehicle repair, laundries and hairdressers. Such services occupy over 100 000 premises; few are equipped with central heating and the scope for energy efficiency is low.

Other Buildings: Some 150 000 premises are included here, the majority of which are places of worship or of entertainment. There are a small number of large buildings in the sector where there is good scope for energy efficiency improvements, but in others the present comfort levels are low and energy consumption may rise as facilities are upgraded.

5.2.3 Public Sector Buildings

National Government: There are 11 000 establishments with some 550 000 civil servants. Plans which called for a 40 per cent reduction in energy demand over a ten year period have been achieved. Many are large establishments and there is still extensive scope for improvements in energy efficiency.

Defence: About 7000 establishments are in this classification for 333 000 personnel, military and civilian. There is a large potential for reduction in energy demand.

Local Government: The wide range of activities includes administration buildings and sport centres but not education. The application of energy efficiency measures has tended to be sporadic and piecemeal, a large improvement potential exists. Leisure centres tend to have reasonably constant demands for both power and heating and so they lend themselves to the introduction of CHP units.

Education: The 42 000 premises included are of widely varying age with as much as 10 per cent of the school stock as temporary accommodation with very poor thermal characteristics. There is scope for saving energy, particularly for space and water heating. However, finance for improvements is limited.

Health: There are about 30 000 health premises in the UK. They range from hospitals to dental surgeries. Hospitals are intensive users of energy with oil being the dominant fuel used. 70 per cent of the buildings are over 60 years old so there is tremendous scope for improvement in energy efficiency.

5.2.4 Recent Energy Trends

The demand for energy in the UK CPBS rose by some 19 per cent over the decade 1975–85, in-line with increases in the indexed service sector output and additional demand for air-conditioning, refrigeration and appliances. Electricity consumption rose by 40 per cent while gas consumption doubled, the latter partly due to gas being substituted for oil and solid fuels in space heating. Over the period 1985 to 1988 total energy consumption fell by some 0.6 per cent; oil and solid fuel consumption fell by 17 and 30 per cent respectively but electricity and gas usage increased by 11 and 6 per cent respectively.

5.2.5 Energy Efficiency

Good housekeeping measures and energy management techniques have the potential to save around 10 per cent of the energy currently used (Table 5.5) A higher potential saving comes from energy efficiency measures and, in terms of minimising CO_2 emission, fuel substitution where appropriate.

Space Heating: The largest potential for saving energy is in space heating, with a potential improvement of up to 50–70 per cent. Most of the savings are by reducing building heat losses but a significant gain is possible through the installation of more efficient heating systems.

The potential saving by the replacement of old boilers is large since over half these in use are over 10 years old and relatively inefficient. Gas condensing boilers are extremely efficient at 85 to 90 per cent. Decentralisation of heating systems in large premises may raise efficiency by reducing transmission heat losses. This may involve the replacement of a large steam boiler by several smaller units located close to points of use. Experience in British Gas has shown that the replacement of large boiler systems by modern appliances generally leads to savings in fuel consumption of over 25 per cent. The modernisation of control systems also has a high potential for energy savings, often as high as 5 to 20 per cent (Playle, 1988).

Passive solar heating can be incorporated into building design to reduce energy demand and the technique is considered to be economic in the UK. Many existing buildings could also benefit from passive solar heating with reasonably short payback periods.

CHP may also contribute to efficient space heating and will be discussed separately.

Water Heating: Substantial savings are possible by reducing water temperature and water use by flow controllers etc. Point of use heating reduces transmission heat losses and is most useful during summer months when large boilers are at low loads and operating at low thermal efficiency.

Air Conditioning: Both the need and energy consumption of air conditioning systems depends largely upon building design. Efficiency can be gained by reducing building heat gain, increasing plant efficiency and by reducing distribution losses. The potential for savings in air conditioning are low and increased use is likely to increase energy consumption.

Lighting: It has been noted earlier that there have been advances in lighting efficiency and availability of the new units. In addition to the availability of more efficient sources better control can reduce unnecessary use. Buildings can also be designed to make maximum use of daylight.

Low energy fluorescent lights are available which use 10 per cent less electricity and all tubes can be fitted with solid-state ballasts which give an efficiency improvement of 25 per cent. Conventional incandescent light bulbs can be replaced by modern fluorescent types which are 75 per cent more efficient. Lighting control systems in large open-plan office buildings have shown savings of over 60 per cent.

Overall the potential energy savings by more efficient lighting methods and control are over 50 per cent. Payback periods for the consumer are generally in the region of 0.7 to 2 years.

Cooking: Better design of kitchens, improved

management and operation together with more efficient appliances and heat recovery could save 20–40 per cent of the energy currently used. Controls tend to be the minimum necessary to comply with safety standards and significant savings could be made by relatively minor changes.

Chemical sterilisation would save much of the energy used in dish washers since temperatures could be reduced.

Refrigeration: Improved insulation, design and controls can reduce energy consumption. Whilst it is difficult to improve the efficiencies of existing equipment, new high efficiency installations offer substantial energy savings. In addition there is the possibility of using built-in heat recovery systems to provide hot water.

Dehumidification: Space heating requirements are reduced by improved insulation and draught prevention but may lead to condensation and dampness. One solution is to install air conditioning or mechanical dehumidifiers but there is a penalty in terms of extra cost and additional energy consumption. Traditional ventilation methods such as fans or grills may avoid condensation but these expel large amounts of warm air and may cause draught. Passive dehumidifiers rely on the generally greater vapour pressure inside a building to transmit it to the outside whilst acting as a trickle ventilator. No energy is consumed and heat losses are minimal. Recent regulations require trickle ventilators in new buildings and they can easily be fitted to existing ones.

5.2.6 Energy Consumption and CO_2 Emission

Overall, CPBS energy consumption of 757 PJ represents 12·7 per cent of the UK total and 14·6 per cent of the total UK CO_2 emission. The latter figure was derived from estimates of energy consumption by fuel type and an appropriate weighing factor for each type to give the CO_2 emission at end use, together with emissions from the generating stations (Table 5.8). No account has been made of CO_2 emissions associated with the production or transport of the fuel used. Thus, in absolute terms, the CO_2 emissions due to the energy demand in the CPBS is underestimated. Suitable weighting factors have not yet been developed to take into account the whole energy cycle.

Projections of national energy usage have been based on the DoEn Energy Paper no 58, submitted to the IPCC (DoEn 1989a), in which the lowest national energy growth rate used is 1 per cent per annum. Since the actual growth rate in energy consumption has in fact been zero per cent over the last 19 years and 0·7 per cent for 1987 to 1988, a rate of 0·5 per cent has also been considered. For a 1 per cent growth rate the CO_2 emission from the CPBS shows a 13 per cent increase by the year 2000 and a 24 per cent increase by the year 2020. The corresponding figures for a 0·5 per cent growth rate are 9 per cent and 17 per cent respectively. Included in these figures is an allowance for a higher-than-average increase in electricity consumption within the CPBS due to the increased use of electrical appliances and air conditioning (Herring et al. 1988).

5.2.7 Energy Savings Using Conventional Technologies in the CPBS

The potential for improved energy efficiency in the sector is estimated to be 40–55 per cent of the total, with the greatest saving in space-heating. With a 55 per cent reduction in energy demand the emission of CO_2 would be reduced by 41 per cent, representing 6 per cent of the total UK emission. Further reductions would be made possible by fuel substitution; for instance, the replacement of electricity for heating and cooking by gas.

5.2.8 New Technologies For Energy Efficiency

The new technologies likely to bring increased energy efficiency include heat pumps and CHP. Both are in limited use and have the potential for

Table 5.8 Energy use and CO_2 emission (as C) in the CPBS (1988)

	Electric	Gas	Oil	Solid	Total
Energy Consumption (PJ)	238	305	170	44	757
C Emission Factor	58·6	14·6	19·8	24·1	
C Emission (000 tonnes)	13 947	4 453	3 366	1 060	22 820
As % of Total	61·1	19·5	14·7	4·6	99·9

substantial energy savings, and thus for reductions in the emissions of CO_2.

Heat Pumps: Heat pumps extract heat from a low-grade source and deliver it elsewhere at a higher temperature. They can perform both a heating and cooling function and are most useful when used in both roles simultaneously. Both electric and gas heat pumps are commercially available, but in CO_2 emission terms an electrically powered heat pump is less desirable than a conventional gas boiler. Gas powered heat pumps, either gas engined or working through an adsorption cycle, may offer significant energy savings (Herring et al. 1988).

Condensing Boilers: Efficiencies of 85–90 per cent are achieved and payback times are generally under 6 years.

Small Scale CHP: In CHP gas and/or oil is used to generate both electricity and heat. They range in size from micro-CHP units with 2 kW output to large units of several MW. They offer efficiencies of around 80 per cent, with about 25 per cent of the fuel converted into electricity and 55 per cent into hot water (Chisholm 1989).

CHP is most efficient for establishments which have a continuous demand for electricity and hot water. Their potential for energy saving is large and the use of CHP is expected to increase substantially in large hotels, swimming pools and leisure centres.

The use of local CHP offers potentially a large reduction in CO_2 emissions to the atmosphere because of the efficient utilisation of waste heat normally associated with electricity generation (DoEn 1989a).

Large Scale CHP: In large scale CHP the CPBS and other buildings in the area are supplied with electricity together with heat in the form of hot water. The amount of CO_2 associated with this energy is reduced since the efficiency of the CHP plant is higher than a conventional power station. There is also the possibility of incorporating combined cycle generation in which the maximum use is made of waste heat to generate electricity during periods when the demand for hot water is low (Playle 1988).

Renewable Energy: Buildings can be designed to make good use of passive solar radiation to subsidise other forms of heating, using large areas of south-facing glass including purpose-built conservatories. Blinds provide temperature control during summer periods. Such measures are economically favourable in the UK with short payback periods. Buildings designed to make good use of daylight can reduce the demand for artificial lighting.

Active solar heating involves the gathering of heat from a collector which is then used for heating and hot water. It has the advantage that the heat can be stored for use when it is most needed. However, payback periods are long and it is not considered economically viable for the UK at current energy prices. Similarly, the conversion of sunlight into electricity by the use of photovoltaic cells is uneconomic. Technically, active solar power does have the potential to reduce energy consumption in the CPBS but its widespread utilisation is probably many years away.

Electrical Appliance Efficiency: With the greatly increased computerisation of modern offices and the extensive use of other equipment, such as photocopiers, the demand for electricity is set to rise. This may be offset by the introduction of more efficient appliances. In the case of computers new low consumption microchips have been developed for battery-powered portables, together with efficient LC displays. Such computers use only a fraction of the power of conventional desktop machines.

Airless Drying: Laundries, hospitals and hotels could exploit the technique of airless drying, on which superheated steam acts as the drying agent, saving the energy normally used to heat ambient air. Savings of up to 90% are possible and the waste heat, in the form of steam, could be used to provide hot water. The technique is considered to be economically viable.

5.2.9 Conclusions

On the basis of 'business as usual', CO_2 emissions associated with energy usage in the CPBS are set to rise between 9 and 17 per cent by the year 2000, with the lower figure being more probable.

The application of conventional energy efficiency technology offers savings of up to 55 per cent, which would reduce CO_2 emissions by 41 per cent. This is equivalent to 6 per cent of the present UK CO_2 emission.

5.3 INDUSTRY SECTOR RESPONSE TO THE GREENHOUSE EFFECT

5.3.1 Introduction

The DoEn estimates that 15–20 per cent of energy utilised in industry can be saved by a combination of good housekeeping and better energy management together with investment in efficiency improvements (with a payback period of 2 years or

Energy usage in the home, commerce and industry, and its effect on the release of greenhouse gases

Table 5.9 Energy efficiency literature

Energy Efficiency Office
DEPARTMENT OF ENERGY

Order Form - Industrial

Information Booklets

Energy Efficiency in Buildings
Advice for Small Firms
Training for Energy Efficiency
Support & Advice for Industry, Commerce & Local Authorities
Guidelines for Local Authority Shared Savings Energy Performance Contracts
Monitoring and Targeting for Energy Efficiency
Energy Efficiency in the Retail Industry
Building on Success (Recommended Practice by The Society of Chief Architects of Local Authorities)
Your Monergy Action Plan for Increased Profit
Role of the Energy Manager

Energy Efficiency in Building Series

1. Schools
2. Catering Establishments
3. Shops
4. Health Care Buildings
5. Further and Higher Education Buildings
6. Offices
7. Sports Centres
8. Libraries, museums, art galleries and churches
9. Hotels
10. High street banks and agencies
11. Entertainment
12. Courts, depots and emergency services
13. Factories and Warehouses

Publicity Material

Posters 'Office'

1. Switch off unwanted lights
2. Turn down radiators
3. Close doors, keep warmth in
4. Close windows, keep warmth in

Posters 'Factory'

5. Switch off when you leave
6. Don't waste steam, report leaks
7. Close factory doors, keep warmth in
8. Don't waste air, report leaks
9. Turn off taps, don't waste hot water

Free posters Order Form

Stickers

When you leave, Switch Off
When you leave, Turn Off
Heating Limit

Journal
Energy Management

Technical Publications

Focus Magazine

1. Focus on Heat Pumps
2. Focus on Food and Drink
3. Focus on Waste as a Fuel
4. Focus on Hospitals
5. Focus on Metals
6. Focus on Finance
7. Focus on Glass, Pottery, Bricks and Cement
8. Local Authorities
9. Focus on the Chemical Industry
10. Focus on Textiles
11. Focus on Lighting
12. Focus on Energy Managers
13. Focus on Paper and Board

Energy Technology Expertise R D

Energy Technology Series

1. Energy Management Systems
3. Energy Efficiency Technologies for Swimming Pools
4. Small Scale Combined Heat & Power
5. Heat Pumps for Heating in Buildings
6. Heat Recovery from High Temperature Gas Streams

Fuel Efficiency Booklets

1. Energy Audits
2. Utilisation of steam for process and heating
3. Economic use of fired space heaters
4. Compressed air and energy use
5. Steam costs and fuel savings
6. Recovery of heat from condensate, flash steam and vapour
7. Degree days
8. The economic thickness of insulation for hot pipes
9. Economic use of electricity
10. Controls and energy savings
11. The economic use of refrigeration plant
12. Energy management and good lighting practices
13. The recovery of waste heat from industrial processes
14. Economic use of oil-fired boiler plant
15. Economic use of gas fired boiler plant
16. Economic thickness of insulation for existing industrial buildings
17. Economic use of coal-fired boiler plant
18. Boiler blowdown
19. Process Plant Insulation
20. Fuel Efficiency in Road Transport

less). As progress is made in energy saving, technology and techniques advance in parallel so that, as a rough guide, the 15-20 per cent energy saving potential appears to remain available at any time. In 1988 the industry sector as a whole was responsible for just over 25 per cent of all UK CO_2 emissions. Readily accessible energy saving measures could yield a reduction in annual CO_2 emissions of the order of 10 million tonnes (as C). In addition, there is a major potential to expand industrial CHP and with it the synergistic development of gas-fired condensing combined cycle electricity generation.

There is much good advice available from the DoEn Energy Efficiency Offices. In Table 5.9 there is a listing of some of the literature available.

In order to assess the potential for improvements in energy efficiency, detailed examination is made of the chemical sector. Chemicals accounted for 20 per cent of the energy consumption of the manufacturing industry in 1988 (DoEn 1989b) and the relative importance is projected to increase significantly (together with papers and man made fibres) while heavy engineering and textiles decline.

5.3.2 Improving Energy Efficiency in the Chemicals Industry

5.3.2.1 The main elements

Over the last 10 years, energy consumption/unit of production in the chemical industry has fallen by 40 per cent, partly by a change in product mix to less energy intensive products, but also as a result of:

a) Design and technology innovation/plant replacement

b) Investment in energy cost saving

c) Monitoring and targeting energy usage

Examples are given from each area in order to provide a basis for extrapolating improvements in energy efficiency from past achievements.

5.3.2.2 Design and technology innovation/plant replacement

A striking example is ICI's leading concept ammonia production process, put into operation in the late 1980's. It has an energy consumption of 28 GJ/tonne. This can be compared with the 39 GJ/tonne for the early natural gas feedstock ammonia process of 1970. The coke-based process which preceeded the natural gas process typically consumed 88 GJ/tonne.

If one were to look towards minimising adverse greenhouse consequences the ultimate ammonia process would perhaps be one based on hydrogen obtained from the electrolysis of water using nuclear electricity. The use of electrolytic hydrogen is not a new concept; it was the hydrogen feedstock for the very early ammonia processes. However, reverting to such a process would be very much more expensive than today's processes which obtain hydrogen from steam and natural gas feedstock and reject the associated CO_2. 'Carbon free' processes, therefore, are not likely to be adopted without world-wide regulatory collaboration and an enormous growth in electricity generated by the use of non-fossil energy.

5.3.2.3 Investment in energy cost-saving

It is generally easier to reduce the demand for process heat than to reduce the demand for electricity. However, some inroads can be made on electrical demand by the use of high efficiency motors, better selection of motor size and variable speed drives.

Process integration techniques are increasingly used to identify opportunities for improving heat integration on existing plants with standard technology. For example, through the use of these techniques, processes with a two year or better payback have been implemented on a vinyl chloride monomer plant to improve energy efficiency from 13·7 to 11·3 GJ/tonne (a 17 per cent reduction) over a five year period. Many of the specific modifications can be replicated on other vinyl chloride plants with comparable benefit.

5.3.2.4 Monitoring and targeting energy usage

As process instrumentation has developed from pneumatic through electronic to today's microprocessor based equipment, the potential for implementing monitoring and targeting and process optimisation has expanded greatly. Systems such as ICI's 'Hamble' and 'Auditor' are examples.

Once again, reductions in the demand for heat are easier to obtain than for electricity. However, we would expect to find savings in electricity consumption arising from monitoring and targeting activities associated with the use of compressed air and refrigeration.

5.3.2.5 Extrapolating improvements into the future

Examining the source of improvements in energy efficiency over recent years does not, in fact, make it any easier to predict the future rate of improvements in the chemical industry, let alone the industrial sector as a whole.

In the chemical sector is seems clear that efficiency will continue to improve but it may be optimistic to assume that improvement will continue at the high rate demonstrated over the past decade.

Estimating future energy demand is extremely complex, being influenced not only by the technological factors considered above but also by the cost of energy itself. In general, projections of industrial energy growth tend to track GNP growth very closely or lag only slightly behind it.

5.3.3 Energy Efficient Technology for Industry

5.3.3.1 Combined heat and power and condensing combined cycle generation

In 1988 the UK had about 1·8 GWe of industrial CHP. Within ICI alone a potential for the installation of 1 GWe of new gas turbine generation CHP has been identified. Given an adequate margin between the price of gas and the value of baseload electricity generation, this is likely to be installed.

If there is sufficient margin between the price of gas and the value of electricity generation to support investment in condensing combined cycle gas turbine generation, such capacity may be added around CHP 'core' generation capacity. This would have the added benefits of providing secure steam-raising capacity and enabling larger, higher efficiency gas turbine units to be used. Preliminary analysis shows that in this way, the initial 1 GWe identified may increase to a total of 3–4 GWe of gas fuelled, gas turbine generation capacity.

Should conventional coal-fired electricity generation, together with heat from gas-fired boilers, be replaced by a gas-fired CHP unit the CO_2 emissions would be reduced by approximately 1 million tonnes (0·27 million tonnes C) for each TWh generated.

Condensing combined cycle gas turbine, when displacing conventional coal-fired units, can reduce CO_2 emissions by approximately 0·55 million tonnes (0·15 million tonnes C) for each TWh of electricity generated. On this basis, 1 GWe of CHP, together with 2–3 GWe of condensing combined cycle, all operating at 80 per cent load factor, would reduce carbon emissions by some 4–5 million tonnes/year.

This potential reduction in carbon emissions from a proportion only of the UK chemical industry is highly significant, equating on current trends to 4–5 years improvement potential for the UK industry sector as a whole.

5.3.3.2 Steam raising plant

Steam accounts for 68 TWh of industrial space heating in the UK but the overall energy efficiency of steam boiler systems is often 40–50 per cent. Replacement by point of use systems, where possible, can raise efficiency to 80 per cent (Playle 1988). New direct heating methods are capable of greatly reducing fuel use; these include high intensity immersion tube burner systems and high temperature direct contact water heaters which are 97 per cent efficient. There are certain applications for steam heating which cannot be easily replaced, such as wood and textile treatment in which moisture, a steady heat flux and limited maximum temperature are important.

5.3.3.3 High temperature heating/melting

Non-ferrous metals are usually heated in crucible furnaces or in top-fired or reverberatory furnaces. Energy consumption can be reduced by direct heating via a ceramic immersion tube, offering efficiencies of 65 per cent (Playle 1988). For processes which require radiant heating, gas-fired ceramic radiant tubes are available which offer lower CO_2 emissions overall compared with electric radiant heating.

5.3.3.4 Heat recovery

In high temperature process heating, the major cause of heat loss when using fossil fuel is found in the flue gases. With a furnace operating at about 1250°C, the flue gases typically leave at 1350°C. This represents a 60 per cent loss of the input energy. Recuperative and regenerative burners have been developed which can recover up to 50 per cent and 90 per cent respectively of the heat in the flue gases.

5.4 THE AGRICULTURE AND FOOD INDUSTRY

Agriculture is a major user of solar energy and the resulting biomass is regarded as a renewable resource. Fossil fuel energy is used to increase the productivity of agriculture and thus to enhance the efficiency of solar radiation capture (Fig. 5.5), often by a considerable amount. This additional support energy is not necessarily to be seen as a net expenditure of energy, especially as some of the biomass may be used as a fuel itself. Biomass is either burnt, if the biomass is dry (e.g. straw, wood), or digested to produce methane, if the biomass is wet (e.g. foliage, excreta). Fermentation to produce ethanol

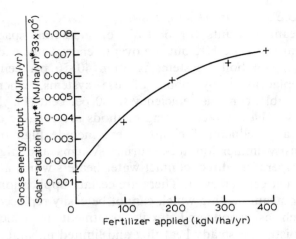

Fig. 5.5. The efficiency of use of solar energy with increasing inputs of support energy.

is also possible but is costly in terms of both money and energy.

The use of support energy varies greatly but Table 5.10 illustrates the national picture.

The usage by the food industry is considerably greater in developed countries as the food industry becomes more sophisticated, whilst farming, under environmental pressures, may become less intensive.

Table 5.11 shows that in farm, i.e. up to the farm gate, by far the largest use of support energy is attributable to the manufacture of nitrogenous fertiliser and the production of feedstuffs.

Energy costs of fertilisers vary with the crop grown and fuel use may be especially high if drying processes are involved (e.g. dried grass).

Animal production for human food is generally less efficient than is crop production, mainly because animals live on crop products and there is thus a further inefficiency of conversion.

In Table 5.12 the efficiency (E) of support energy use is expressed as units of energy produced in the product; E tends to exceed 1·0 for crops and rarely to exceed 0·5 for animals.

Table 5.10 Primary energy involved in food production, UK, 1973

	Fossil Fuel Energy (PJ)	% of National Consumption (9 260 PJ)
Agriculture (to farm gate)	361	3·9
Processing, packaging and distribution	648	7·0
Food storage and preparation	449	4·9
Total	1 458	15·8

Source: White (1974).

Table 5.11 Use of support energy within production systems

	Winter Wheat	Potatoes
(a) Crops GJ/ha/year[a]		
Fertilizer N	10·40	14·0
P	0·70	2·45
K	0·45	2·45
Tractor field work	3·24	3·99
Other field work	1·29	10·08
Sprays	0·40	1·24
Drying fuels	1·72	–
Drying machinery	0·57	–
Seed shed fuels/storage	–	2·14
(b) Animals (MJ/Progeny reared to slaughter)[b]	Poultry	Rabbits
Number of progeny/year	100	80
Feed	Commercial mix	50% Barley/50% Dried
Energy inputs to feed	112·73	79·54
Capital	16·95	10·76
Electricity	13·91	5·59
Fuel for heating	3·71	10·45
Tractor fuel	1·47	1·54
Field machinery	0·91	0·70
Veterinary and medical	1·83	0·71
Water	0·13	0·12
Litter	0·27	

[a] Leach, 1975.
[b] Spedding et al. 1976.

Table 5.12 Energetic efficiencies (E) in agriculture (E = gross energy in food produced/support energy used to produce it)

Product	E
Milk	0·33–0·62
Beef	c. 0·18
Broilers	c. 0·1
Eggs	c. 0·16
Wheat	2·2–4·6
Barley	c. 1·8
Potatoes	1·0–3·5

Source: Spedding (1982).

Table 5.13 Efficiency of energy using food production, UK, 1973

	MJ of energy in product per MJ support energy used
Wheat at farm gate	3·200
Bread—white, sliced, wrapped	0·500
Milk at farm gate	0·650
Milk bottled and delivered	0·595

Source: Leach (1976); Spedding and Walsingham (1976).

However, since even more support energy is used in the food industry, much depends upon the extent to which crop and animal products are processed. The examples of bread production from wheat (by grinding etc.) and the delivery of milk, given in Table 5.13, show that the differences in crop/animal production efficiencies can be reversed during processing, leaving the overall efficiencies little different.

There is not, to our knowledge, a comprehensive study available on the potential energy savings in agriculture. However there clearly are savings to be made. Energy efficient crop dryers use 20–50% less energy than existing units. Improvements in insulation and heating systems could save more than 20% of the energy used in greenhouses, and recent developments in the use of widespan gantries for cultivation have led to energy savings of 75%. The potential savings in agriculture through increased efficiency is likely to be 20% or more (Day, Pers. Comm.). However, resistance to innovation, high capital costs or poor payback periods may limit the rate of implementation of new technology.

REFERENCES

CHISHOLM, D. W., (1989). *Opportunities for gas in the developing commercial market*, Paper presented at the 126th Annual General Meeting of the Institution of Gas Engineers.
DAY, W. (Pers. Comm.). Private communication to the Working Party 1990.
DoEn, (1989a). *An evaluation of energy related greenhouse gas emissions and measures to ameliorate them.* Intergovernmental Panel on Climate Change, Department of Energy Paper No 58.
DoEn, (1989b). *Digest of UK Energy Statistics 1989*, HMSO.
HERRING, H., HARDCASTLE, R. & PHILLIPSON, R. (1988). *Energy use and energy efficiency in UK commercial and public buildings up to year 2000*, ETSU DoE Energy Efficiency Series no 6, HMSO.
JOHANSSON, J. B. et al. (1989). *Electricity—Efficient Energy Use and New Generation Technology and the Planning Implications*, Lund University Press, Sweden.
LEACH, G., (1975). *Energy and Food Production*. Int. Inst. for Environment and Development, London.
LEACH, G., (1976). *Energy and Food Production*, IPC, Science and Technology Press (London).
LOVINS, A. et al. (1989). *Least Cost Energy For Solving the CO_2 Problem*, Rocky Mountain Institute.
PLAYLE, B. A., (1988). *Exploitation of developments in utilisation in the non-domestic market*, 54th Autumn Meeting of the Institute of Gas Engineers.
SHORROCK, L. D. & HENDERSON, G. (1990). *Energy Use in Buildings v Carbon Dioxide Emissions*, BRE Report 170.
SPEDDING, C. R. W. (1982). *The comparative economics of agricultural systems*. In: *Energy and Effort*, Ed. G. A. Harrison, Symp. of the Soc. for the Study of Human Biology, Vol. 22.
SPEDDING, C. R. W. (1984). *Energy use in the food chain*. Span, 27(3) 116–118.
SPEDDING, C. R. W. & WALSINGHAM, J. M. (1976). *The production and use of energy in agriculture*. J. Agric. Econ, 27, 1, 19–30.
SPEDDING, C. R. W., WALSINGHAM, J. M. & BATHER, D. M. (1976). *Alternative animal species for more efficient farming*. In: BATHER, D. M. & DAY, H. I. (eds.) *Energy Use in British Agriculture*, Reading University Agric. Club.
WHITE, D. J., (1974). *Prospects for great efficiency in the use of different energy sources*. Phil. Trans. R. Soc. Lond. B., 281, 261–75.

Section 6

Energy Usage in Transportation

6.1 INTRODUCTION

Demand for transport has grown dramatically throughout the 20th Century, and is inextricably entwined with the process of economic growth. Similarly private car ownership has come to be identified as one of the justly earned fruits of economic growth, and has expanded accordingly.

The degree of freedom granted by the ownership of a car has a near symbolic importance in the late twentieth century. At the same time air travel has become almost universally accessible in the developed world, and has played a principal role in the development of international trade.

The purpose of this chapter is to examine the potential for the long term reduction of greenhouse gases resulting from the transportation sector. It has to be appreciated however that changes to fuel quality or modifications to vehicles which are required for other health or environmental reasons can significantly affect the volume of greenhouse gases. For example the removal of lead from gasoline and the use of catalytic convertors for excellent reasons, nevertheless have increased refinery and car operation energy requirements and therefore the emission of greenhouse gases.

6.2 THE CURRENT SITUATION

Since the oil price shocks of 1973 and 1979 demand for oil has contracted in almost every sector except transportation where it has increased inexorably.

By 1988 transport represented 29·5% of total delivered energy in the UK, and 58% of total oil demand. As a percentage of oil demand for energy purposes alone (excluding bitumen, plastics, lubricants etc.), transport represents 71·5%. In 1986 transport energy demand overtook industrial demand for the first time.

In 1988 £16·5 billion was spent on energy for transportation, representing 42% of the UK's total expenditure on energy. These statistics indicate the growing dependency of the oil industry on the transport market and, in turn, the capacity for growth that the sector has shown.

Fuel demand from the road haulage sector grows proportionally with the overall level of economic activity in the country—the activity of the vehicle stock has a considerable degree of elasticity, but ultimately vehicle numbers will expand and contract according to demand for haulage services.

However fuel demand for private transport seems to be limited only by a notional saturation point for vehicle ownership. That may be the point at which every person eligible to drive owns at least one vehicle, or it may be a point at which other factors

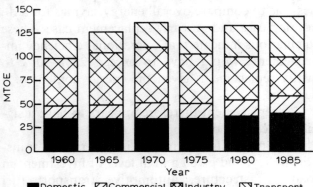

Fig. 6.1. UK final energy consumption by sector.

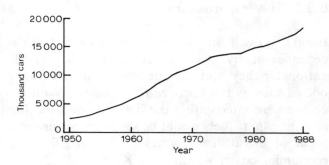

Fig. 6.2. UK car stock 1950–1988.

such as road congestion conspire to limit further growth in demand for cars.

The UK is now widely regarded as approaching a road congestion crisis. Although there is still a considerable number of people who do not own cars but have the potential to do so, the infrastructure is struggling to support the existing road vehicle stock. Nonetheless in 1989 new car registrations once again broke all records, growing by 4·2% over 1988. In 1988 private and business cars consumed an estimated 50% of all transport energy.

Air travel has grown faster than other modes of transportation with UK airlines increasing from 9·8 billion passenger miles in 1963 to 86 billion in 1986, about 10%/year increase.

In the UK aircraft operate with high load factors mainly as a result of a large proportion of non-scheduled tour operations. In 1986 their operations had an average load factor of 92%.

Estimates vary as to the precise contribution made by transport to carbon dioxide production in the UK. However the Digest of Environmental Protection and Water Statistics (DoE 1989) suggests that UK carbon dioxide emissions from petroleum used in road transport amounted to over 100 million tonnes in 1987. This represents about 18% of total UK carbon dioxide emissions from fossil fuels. If aviation is included this rises to about 22%.

6.3 UK GROWTH IN TRANSPORT FUELS

The Department of Transport forecasts that road vehicle miles will grow by between 83 and 142% by the year 2025. The Department's forecasting record is not good, and it might be said that this apocalyptic vision is in reaction to years of underestimating the rate of road traffic growth. The Department of Transport has produced a White Paper, 'Roads to Prosperity' (DoT 1989), which proposes satisfying this demand with £6600 million's worth of road infrastructure investment over the next ten years.

6.3.1 The Parameters

Future growth in demand for transport fuels depends largely upon government policy—both national policy and the dictates of international bodies such as the European Parliament. Without government intervention the limits on demand for road transport fuels would be set by the saturation point of private car ownership and the level of economic activity within the UK.

Based on current demographic data one car to

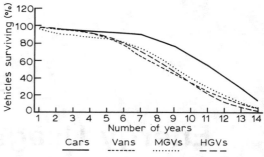

Fig. 6.3. Vehicle survival rates.

one potential license holder would amount to a fleet of roughly 35 million private cars, compared with the 18·4 million registered in 1988. Even with a laissez-faire Governmental approach it seems likely that congestion will limit the growth of vehicle ownership before that point. There is much to suggest that additional road infrastructure would tend to release existing pent-up demand.

During the 1980s advances in automotive vehicle technology continued to produce fuel savings. However the cheapening of the relative cost of fuel has been followed by a trend towards an increased emphasis on power and speed, creating a situation where the car stock is significantly less fuel efficient than technology might allow. If attitudes were to change, the rate of improvement in overall fleet efficiency would be constrained by the rate of vehicle replacement which currently stands at around 8% per annum.

6.3.2 Car Replacement

With more efficient engines becoming available, the possibility of early replacement of older cars to obtain benefits of the lower fuel consumption has to be considered.

Using data from a 1974 Motor Industries Association report which analysed car energy requirements for manufacturing and operation, it is possible to compare overall energy savings for a do nothing case and an early replacement case. These calculations indicate that operational efficiency improvements would have to exceed 25%–30% to justify replacement, given the energy content of a new car.

Car replacement therefore appears to have a significant carbon dioxide emission cost.

Without significant changes in government transport policy, and given a high level of investment in road infrastructure, automotive transport fuel demand might continue to follow a path close to the one indicated in Fig. 6.4.

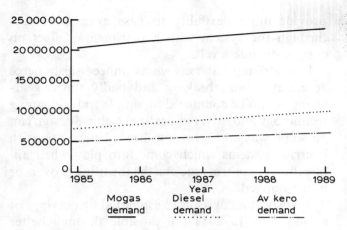

Fig. 6.4. Transport fuel demand.

An alternative forecast assumes that congestion of the roads will affect car sales and although there will continue to be some increase in transportation fuel demand, a flattening of the curve will occur.

The increased demand for transport fuels will not be restricted to the UK and worldwide demand for fuels will produce a natural supply-demand effect of increased crude costs as the availability declines. Of particular importance is the development of the market in a revitalised Eastern Europe and important in the long term is what happens in large population centres such as China. The increase in prices even with the current level of taxation will provide an incentive to use fuels more efficiently, thereby reducing future demand below forecast levels.

6.4 IMPLICATIONS

A World Wide Fund for Nature report published in November 1989 forecast that, based upon Department of Transport upper range estimates, carbon dioxide emissions from the road transport sector could double by the year 2020. The report suggests that even given a one percent per annum improvement in road vehicle fleet efficiency from the early 1990s, combined with the Department of Transport's lower range forecast, carbon dioxide emissions would increase 20% by the year 2020 (DoT 1989).

The World Climatic Programme in Toronto 1988 recommended the target of a 20% reduction in world carbon dioxide emissions by 2005. A conference of non-government organisations in Hamburg the same year stated that to achieve this objective developed countries would have to reduce their carbon dioxide output by a significantly greater percentage. The proposal suggested that the UK would have to achieve a 30% reduction in carbon dioxide emissions by the year 2000 and 60% by the year 2015.

6.5 GENERAL DISCUSSION ON POSSIBLE FUTURE STRATEGY

The analysis of the UK transport sector's contribution suggests that ways should be found to reduce transport-related carbon dioxide emissions. Evaluation of the various options for achieving this objective is complex and hampered in many instances by the lack of reliable data.

Other more localised environmental issues are liable to become entwined with the central aim of reducing carbon dioxide emissions, and none more so than the question of road congestion. Government policy makers will find themselves confronting questions of personal liberty should they attempt to limit the use of cars by road pricing or any similarly direct means.

Alternatively a range of technical opportunities exists for improving the efficiency of road vehicles. However it is questionable whether the required reductions in emissions can be achieved by this route if vehicle numbers continue to rise.

6.6 TECHNICAL

6.6.1 Engine Technology

Current engine technology has the potential for efficiency gains if fuel economy were to have a higher priority than power in the minds of vehicle purchasers.

In addition to possible developments of the conventional spark ignition engine, the development of new more efficient engines should be encouraged. For example the two-stroke orbital engine developed by Mr Ralph Sarich has recently entered production under the auspices of General Motors. This engine is claimed to better the power to weight advantage of a conventional two-stroke engine and to have fuel efficiency gains of 25–30% relative to current four-stroke technology. With regard to other emissions nitrogen oxides are claimed to be reduced by 80–90%, hydrocarbons by 50% and carbon monoxide by 90% compared with conventional four-strokes. The reduced weight and size of the unit contributes significantly to these gains. The engine is apparently 25% cheaper to make on account of its relative simplicity. General Motors will be using the engine in small cars within two years, and in marine outboard motors at an earlier date.

In many respects the significance of these developments is the challenge they may represent to diesel's fuel efficiency advantage. Development of diesel engines for cars has been relatively neglected for decades. If fuel efficiency were to become the main technical drive in the motor industry, the development of diesel engines is likely to accelerate.

The flexibility within the refinery sector to provide more diesel fuel at the expense of petrol production is determined by the type of crude available for processing and the upgrading facilities available. Although it is true that further investment could be made there are limitations caused by the natural carbon to hydrogen ratio of the products and the crude processed. Hydrogen availability in the refinery could limit future upgrading. Although hydrogen can be manufactured, this itself releases carbon dioxide. The costs of increasing diesel fuel output could be disproportionately high both in capital and in energy.

Part of the diesel efficiency advantage comes from the higher gravity and higher carbon to hydrogen ratio of diesel versus gasoline. The higher carbon content of diesel would therefore offset some of the efficiency benefit. A diesel engine emits about 10% more carbon dioxide than a gasoline engine if they each have the same fuel consumption expressed in terms of miles per gallon.

Progressive penetration of the car stock by electronic engine-management and injection technology should improve fuel efficiency to some extent.

6.6.2 Traffic Management

A number of new technologies for the improvement of traffic management have emerged over the last two years. Simple technologies, such as electronic monitoring of traffic lights to prevent violation of signals, will have a beneficial effect on traffic flows. Other more advanced systems such as Autoguide —an in-vehicle navigation system utilising roadside beacons to inform drivers of their precise position —will reduce time and fuel spent by drivers getting lost or utilising inefficient routes. The system will eventually be able to warn of obstructions or traffic congestion and advise of alternative routes thus generally smoothing the flow of vehicles. However the unblocking could alternatively be used to provide more flexibility to pass even more cars through the system with a detrimental affect on carbon dioxide levels.

Poor driving habits, such as unnecessarily fierce acceleration and braking and badly timed gear-changes, can be countered by simple traffic-calming technology and driver education. Vehicle design can go some way to correcting driver behaviour, with override systems which come into play when application of controls such as the throttle are over- or under-applied.

Transport analysts have estimated fuel savings of as much as 15–20% are possible through better traffic management and improved driving technique. It has been estimated that a reduction in average motorway speeds from 120 km/h to 90 km/h could result in fuel savings of as much as 10–30% per car, dependent mainly on engine size.

6.6.3 Other Vehicle Efficiency Improvements

A recent ETSU report (Martin & Shock) suggests that on average a 10% reduction in vehicle weight results in a 3–5% efficiency improvement, although efficiency gains in this area may be limited by the conflict between vehicle weight reduction and safety requirements. However the introduction of new materials such as carbon fibre, plastics and ceramics and increased use of glass-fibre and aluminium might allow further weight reduction without safety compromises.

The afore-mentioned study proposes that improvements in gearbox, technology including the use of continuously variable transmissions offer potential gains of up to 15%. Automatic gearboxes which cancel out poor driving technique, and wider ratio manual gearboxes with reduced mechanical drag, all have energy efficiency implications. Martin and Shock calculate that the continual improvement in vehicle aerodynamics could result in short-term efficiency gains of between 3–5% as newer vehicles penetrate the stock with longer term gains of as much as 10–20% possible.

After levelling off during the recession of the early 1980s the average size of car engines began to rise again in 1986. By 1988 the average engine size was 1500 cc, higher than ever before, while 8% of cars had engines of over 2000 cc. The increase in average engine size is a clear manifestation of the desire for power and increased car size at the expense of economy that had characterised the new car market in the late 1980s. Larger cars will normally have

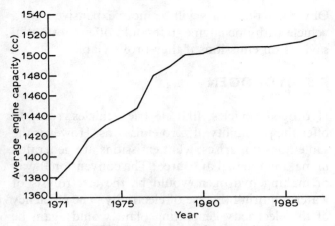

Fig. 6.5. Average car engine capacity.

bigger engines, higher body weights, greater drag factors and thus higher petrol consumption.

6.7 ALTERNATIVE FUELS

Some of the most widely discussed alternative transport fuel options may offer benefits in terms of reducing emissions of hydrocarbons, oxides of nitrogen and sulphur, and carbon monoxide. These pollutants are harmful particularly as contributors to the problem of acid rain. In terms of reducing emissions of carbon dioxide alternative fuels must compete with the traditional transport fuels in terms of overall thermal efficiency.

6.7.1 Alcohols

Methanol and ethanol have certain technical problems as replacements for gasoline or diesel. In the first instance they are corrosive to some elastomers and certain metals used in modern motor vehicles. They also fail to provide the engine lubricating properties that both diesel and gasoline offer in some measure. These are problems which can be overcome by technical development. The combustion of alcohols also tends to produce aldehyde emissions, requiring the use of a catalytic converter to clean the exhaust.

In order to offer the same range as a petrol vehicle between fuel stops an ethanol fuelled vehicle has to carry one and half times the volume of fuel. Since ethanol weighs about the same per litre as petrol this represents a 50% weight penalty plus reduced space in the vehicle. Methanol requires twice the amount of fuel to cover the same distance as a petrol vehicle, with a 100% weight penalty.

The need to carry extra weight in fuel reduces the efficiency advantage of alcohol fuels to some extent, however the carbon dioxide emissions from the combustion of alcohols are less than those of the equivalent petrol engine. However the method used in the production of alcohol fuels is fundamental to their efficiency in reducing carbon dioxide emissions overall.

Ethanol produced from the fermentation process has a potential to provide transportation energy without a net contribution to the greenhouse gases. However the availability of suitable land for biomass is limited and ethanol must compete with other fuels on an efficiency basis.

On balance other forms of biomass appeared more attractive than ethanol production. The low concentration ethanol produced from the fermentation process requires energy to upgrade to motor fuel quality (98%) and experience with the process shows it to be very expensive relative to current fuels. In addition special attention to the method of storage and distribution to prevent water ingress would be required, particularly in countries with a persistent rainfall.

Methanol produced by the oxidation of natural gas will offer little improvement on diesel in terms of carbon dioxide emissions.

6.7.2 Gases

Precedents exist in countries such as the Netherlands and New Zealand for the use of liquefied petroleum gas and compressed natural gas as automotive fuels. LPG is a product of crude production and refining, and consists primarily of propane with some butane, whilst CNG being simply natural gas is composed mainly of methane.

Both gases are compatible with existing spark-ignition (petrol) engines, after minor modifications. However both gases need to be stored in tanks under pressure—LPG requires up to 200 psi pressure, whilst CNG has to be stored at 2400 psi to remain liquid. These tanks are necessarily larger and heavier than conventional petrol tanks—in the case of CNG tanks, up to one hundred times heavier. In countries where LPG and CNG are currently in use, lack of a nationwide refuelling network requires gas fuelled vehicles to be able to use petrol as well. Consequently the vehicles in question are rarely tuned to run at optimum efficiency on gas.

Where gas fuelled vehicles are fully optimised for gas, fuel savings of 15% for LPG, and 10–15% for CNG can be expected compared with the equivalent petrol vehicles. The lesser efficiency of CNG can largely be attributed to the extra weight of the gas cylinder.

Carbon dioxide emissions from these vehicles are between 25 and 40% less than for petrol vehicles. However atmospheric emissions of propane, butane and methane can currently result from refuelling and leaks may well counteract the reductions in carbon dioxide. Improvements in the methods of distribution to minimise leakage are required to obtain full benefits. However the availability of LPG is limited and should not be considered as a replacement for petrol. It may best be used in a particular sector by adjusting taxation to make it attractive. For example taxis or buses operating within city limits on LPG could reduce pollution and would avoid the necessity of providing an expensive nationwide supply infrastructure as would be required for more general sales.

6.7.3 Electricity

There has been a great deal of discussion of the problems of storing electricity for automotive applications in a way that improves on the performance of the lead acid battery.

During the course of 1989 a number of manufacturers produced prototype electric cars with significantly improved performance and range compared with what had been previously been believed possible. This culminated with the unveiling of General Motors' 'Impact' prototype. This offered a claimed 0–60 mph time of 8 seconds and a top speed of more than 100 MPH. A range of 124 miles was also claimed for the vehicle. These claims add up to a car which in performance terms, if not range, can compete with combustion engined cars.

However any achievements in the development of electric cars will mean nothing if the method for generating the electricity with which they are charged, including electricity transmission and charging losses, does not result in a reduction of carbon dioxide emissions versus petrol engined vehicles. To have a significant effect on greenhouse gases, the move to electric cars is dependent on large scale incremental production of electricity with low greenhouse gas emissions.

Certain manufacturers such as Volkswagen have experimented with hybrid vehicles. These use an electric motor in urban conditions where a continuously running internal combustion engine is highly inefficient and polluting. On open roads and above certain speeds they then revert to an internal combustion engine which recharges the batteries.

Obviously these cars will be more expensive than a vehicle with one engine, but could offer substantial savings on emissions if they prove viable.

6.8 HYDROGEN

Hydrogen vehicles, like electric vehicles, seem to offer the possibility of zero emissions. However the same question arises: what emissions are generated in making the fuel at source? The conventional way of making hydrogen would be the electrolysis of water requiring the use of electricity. The efficiency of the electricity generating plant would again be the stumbling block in attempting to alleviate greenhouse gas emissions.

Storage of hydrogen for automotive purposes as metal hydrides has an energy density half that of gasoline, but nonetheless is still manageable and no worse than that of methanol.

6.9 GOVERNMENT POLICY

The pace of implementation of the various options for reducing carbon dioxide emissions will be largely determined by the steps taken by governments to encourage change. Every option will involve additional costs for the private motorist, and many will be inconvenient or appear to offer inferior transport. The UK Government's use of a tax incentive to accelerate the uptake of unleaded petrol during 1989 is a measure of how government intervention can succeed. The US Government's so-called CAFE, or corporate average fuel efficiency regulations, have been instrumental in reducing the average engine size of American cars. The code decrees that a car manufacturer's model range—weighted for the number of each model produced—must average 26 miles per US gallon. In order to accelerate the introduction of some of the innovations detailed above, government intervention may be essential.

6.10 PLANNING POLICY

The Government of the Netherlands have suggested a number of innovative ways in which government intervention might help to reduce vehicle emissions. They suggested that planning regulations might force new businesses to locate close to the residential centres from which their workforce is drawn. The report also recommends that businesses be required by government to meet transport-use reduction targets each year, which would include the mileage travelled to work by their workforce. This it was suggested might result in employers providing buses to work, organising car pooling and collaborating with local public transport organisations. In addition reducing transport use would become an important business objective rather than simply an operational cost.

In the UK subsidies for company cars stimulate growth in the vehicle stock. The removal of the so-called perk car has been widely recommended.

6.11 AIR TRAVEL

Air travel energy costs have reduced significantly since 1963 with a specific energy reduction from 9·3 to 2·5 MJ/passenger kilometer by 1980, after which it levelled off. Part of this reduction must result from increased aircraft engine efficiency and size, but the major change has been brought about by the increase in tour operations over this period. These have increased by about 19%/year against an increase of 8·5% in scheduled air travel.

As with road transportation the congestion of air lanes should cause a movement to fewer flights of larger aircraft on the shorter routes leading to a reduction of the energy cost per passenger kilometer.

6.12 PUBLIC TRANSPORT

UK Government policy with regard to public transport throughout the 1980s has stressed the need for public transport systems to be profitable. In the case of rail, this has meant that all new investment in infrastructure has had to be justified strictly in terms of return on capital employed. The result of this policy has been that the UK rail network is much inferior to those of most North European countries. In the UK rail accounts for less than 8% of total passenger kilometres travelled, and less than 9% of freight tonne kilometres. Rail's low share of total passenger kilometres is a result of competing with the private car—and this in turn has meant that much new investment has not been justifiable on the basis of potential profitability.

If rail is to become a viable alternative to the private car the capacity of the network has to be dramatically increased. The timescales needed to create a rail system capable of supplying even half of the demand currently supported by the road infrastructure, would run into several decades.

Bus systems could more readily satisfy a rapid growth in demand for public transport, but government policy would need to raise the costs of private car use to a high level in order to encourage a transfer between private and public transport. The benefits in terms of overall reductions in carbon dioxide emissions obviously depend upon the degree of transference, but it has been estimated that 10% transfer from private to public transport would result in a 6% reduction in energy demand.

6.13 WATER BORNE TRANSPORT

Developments of internal road and rail links, combined with the greater use of pipelines by industry, has reduced demand for marine freight in the UK to some extent. However in 1986 despite the decline in both national and international shipping, 83% of visible international trade in value and 99% in weight was transported by sea.

A shift in energy use for shipping has occurred over the last twenty years as fuel oil-driven steam turbines have given way to marine diesel engines. At the same time an increasing proportion of marine energy demand has come from subsidiary systems such as refrigerators.

Data for the marine sector is necessarily fairly vague since both UK and foreign vessels refuel within and outside the UK depending on fuel price. Therefore the specific energy consumption by UK vessels, or vessels visiting the UK, is not certain. In addition much of that energy consumption occurs in international waters, and therefore measures to monitor or control energy consumption will be difficult to formulate and even more difficult to enforce.

6.14 SUMMARY

The capacity for significant reductions in transport energy demand exists even with current technology. A rationalisation of the use of private cars—reducing wasteful and unnecessary journeys and increasing the percentage if not the number of longer fuel efficient car journeys, greater use of public transport, and the penetration of energy efficient technologies—all require price incentives. A balance of price incentives and planning restrictions with due regard for the poorer sections of the population may be the best way to encourage judicious use of the private car and truly reflect the environmental and social cost.

REFERENCES

DoE, (1989). *Digest of Environmental Protection and Water Statistics*, HMSO.

D o Transport, (1989). *Roads for Prosperity*, HMSO.

World Wide Fund for Nature, (1989). *Atmospheric Emissions from the Use of Transport in the UK*, WWFN.

MARTIN, D. J. & SHOCK, R. A. W. *Energy Use and Energy Efficiency in UK Transport up to Year 2101*, ETSU.

Government of the Netherlands, *To Choose or to Lose, National Environmental Policy Plan 1988-1989*.

Section 7

Non Energy Related Sources and Sinks for the Greenhouse Gases in the UK

Chapters 4 to 6 have considered the release in the UK of greenhouse gases as a result of the production, conversion and use of energy in all its forms. For completeness, this chapter considers those sources of the greenhouse gases which are not released naturally and which are not directly related to energy. The one possible way by which one might remove CO_2 from the atmosphere; that is by planting vegetation and in particular, trees is also considered.

There is little doubt that, of these non-energy sources, the release of the chlorofluorocarbons (or CFC's) is of most importance.

7.1 CFC'S AND THEIR SUBSTITUTES

The chlorofluorocarbons, also known as freons, are man-made compounds which, as they are non-toxic and inert, are suitable for use as propellants in aerosols, as the working fluid in refrigerators and air conditioning plant, as solvents and as foam producing agents. About 350 thousand tonnes of CFC 11 ($CFCl_3$) and a similar quantity of CFC 12 (CF_2Cl_2) have recently been produced each year together with smaller amounts of other CFC's. Much of these gases finish up, sooner or later, in the atmosphere where they persist for the order of 100 years. Strong indications were discovered in the late 1980's that these gases were damaging the ozone layer which protects the earth from harmful ultra-violet radiation, causing a hole to form in this layer over the Antartic. A major international effort was mounted to limit the use of these substances worldwide. The Montreal Protocol, resulting from the Vienna Convention of 1985, set a goal of a reduction, relative to 1986 levels, in the production and consumption of CFC's of 50% by 1999.

Besides their effect on the ozone layer, these gases are also powerful greenhouse gases. While the concentration of the CFC's (of the order of 2·8 ppbv) is 125 000 times smaller than that of CO_2 (DoE 1989) their effect on infra-red radiation is 10 000 to 20 000 stronger per molecule of gas (as discussed in Chapter 2) and hence they are at present responsible for about 20 per cent of the total greenhouse effect. Their release is world-wide and is such that, without any limitation being put on their use, they would be responsible for an estimated 30 per cent of the total greenhouse effect by the year 2030. The Montreal Protocol aims, at present, at only a 50 per cent cut in production but the UK is advocating a total ban by the end of the century.

It has been suggested that, in the UK (and elsewhere) there could be savings of about 30 per cent in the total quantity of CFC's or their substitutes used that can be achieved by better husbandry of these fluids (McCulloch 1990). More care can be taken to reduce losses from plant and the fluid in obsolete or redundant refrigerators and air conditioning plant should be recovered and recycled. Of the remaining 70 per cent, about 30 per cent could be taken up by gases such as ammonia which will not add to the greenhouse gases. Fluorocarbons could be used to meet the rest of the demand. Those chemicals have a lower ozone depleting contribution and would make a much smaller contribution to the greenhouse effect.

Thus, HFC 134a (1,1,1,2-tetra-fluoroethane) or HCFC 22 (chlorodifluoromethane) could replace CFC 12 in refrigeration and air conditioning units. HFC134a has a reference lifetime in the atmosphere of 15·5 years compared with 120 years for CFC 12 with an effect on thermal warming per molecule only 0·085 times that of CFC 12. The equivalent figures for HCFC 22 are 15·3 years and 0·11. HCFC 123 (1,1,1-trifluorodichloroethane) which can replace CFC 11 in foam blowing has a reference life of only 1·6 years compared with 60 years for CFC 11 and has only 0·1 times the effect on thermal warming. HCFC 141b (1,1-dichloro-1-fluoroethane) which may be used as a solvent to replace CFC 113 has a lifetime of 7·8 years compared with 9·0 years for CFC 113 but its effect on thermal warming is only 0·07 times that of CFC 113.

There is thus the possibility of reducing the release of new CFC's and using substitutes so that the additional contribution from this source to the greenhouse effect will be much smaller than today.

This will depend on:

(1) ensuring that CFC's at present sealed in refrigeration and air conditioning plant are recovered and not left to leak into the environment.
(2) that CFC's are replaced in all applications in which they are now used by replacement compounds which, as a result of technological testing, are shown to be environmentally acceptable and which will not make a significant contribution to the greenhouse effect.

7.2 HYDROCARBONS AND SOLVENTS

There is no doubt that many hydrocarbons and solvents used in industry give off some methane though there is no immediate evidence that the quantities are significant. The life-time of the higher hydrocarbons in the atmosphere is likely to be so short as to make their influence unimportant. The Working Party has not, however, considered those in detail and in particular the possibility that they may be significant precursors of tropospheric ozone, needs further consideration.

7.3 AGRICULTURE

The two main sources of greenhouse gases from agriculture on the global scale occur as emissions of methane; on one hand from ruminant farm animals and, on the other, from rice or paddy fields. Both of these are discussed in more detail in Chapter 8 where the position outside the UK is considered. As far as agriculture in the UK is concerned, however, rice fields are obviously not a consideration. While some reduction in the number of beef cattle and sheep could be envisaged as a result of a major effort to change the eating habits of the populace, the number of animals is small compared with the rest of the world.

Agricultural activity is, to a large extent, neutral in terms of the nett exchange of carbon dioxide. In crop production, the CO_2 taken out of the atmosphere by the growing crops is released on decay, combustion or digestion of the biomass produced. Only where the release is delayed for a number of years; as in trees considered in the next section or with the small quantities of, for example, straw used for construction purposes: is there a significant delay in the release of the carbon. In general the carbon in the biomass is released as CO_2 though microbial activity in decaying vegetables may give rise to methane which has, as discussed in Chapter 2, a larger input to greenhouse warming than CO_2.

Agriculture can increase the natural release of nitrous oxide from the soil. This release is mainly the result of microbial processes in soil and water, both nitrification and denitrification. Both land cultivation and the application of nitrogenous fertilisers can increase the natural effect by an amount dependent on the method of application and type of fertiliser. The emission attributable to cultivation including the use of fertilisers is about the same as that estimated to arise from oceans and freshwater sources and about half that naturally arising from the soil.

7.4 FORESTRY

The area of productive woodland in the UK has been increasing steadily since 1950, generally at about 20 000 to 30 000 hectares per year. It has now reached 2·1 million hectares (approaching 10 per cent of the total land mass). The larger part of this woodland is Sitka spruce which is now in an exponential phase of growth. The annual productivity of stem wood was about 4·6 million cu m per year in 1980 and is expected to double by the year 2005. The present (1990) figure of 6·0 million cu m per year is equivalent to an annual growth in the total carbon stored of about 2·0 million tonnes.

This figure is likely to reach 2·8 million tonnes per year in 2000 as shown in Table 7.1 (Jarvis 1989, Forestry Commission 1988).

This increase will be as a result of both a continuing increase in area of forest estate and of increases

Table 7.1 Annual sequestration of carbon by UK forests

Year	million tonnes/year
1980	1·5
1990	2·0
2000	2·8

in the rate of incorporation of carbon. Increase in the latter will result particularly from the use of better, ex-agricultural soils for a proportion of future planting, more intensive fertilisation and better silvicultural management. However, the rate of increase in the incorporation of carbon may slow down as a result of a swing towards the planting of a higher proportion of slower growing hardwoods in the next century.

Carbon is removed from the atmosphere and is locked up in the accumulating mass of trees. When the trees are harvested, much of that carbon is immediately at risk of oxidation and return to the atmosphere. With conifers, this may occur on a cycle of approximately 60 years depending on species, site, market demand etc. From a long term perspective, hardwoods on, say, a 200 year rotation may be more effective in sequestering CO_2, although much slower growing, because the risk of oxidation and return to the atmosphere occurs less frequently and the manufactured products tend to be longer lasting.

Clearly, the more uses and, hence markets for long-lasting wood products there are, the less the amount of harvested wood that will be rapidly converted to CO_2. The burning of woody biomass or waste to produce useful energy in the form of steam or electricity as a substitute for coal or oil should also be considered seriously.

Provided that the annual cut does not exceed the annual growth increment, so that the system is neutral with respect to land use, biomass and CO_2, then wood from industrial plantations is a sustainable source of energy that has much to recommend it.

However, to put this potential for influencing CO_2 emission into perspective, let it be assumed that a reasonable target to set is a doubling of the area of woodland in the UK, so that approximately 20% of the total land mass is wooded. This area of woodland would provide a sink for some 4 million tonnes of carbon per annum, assuming that all the trees are in the required stage of growth. This is of the order of 2·5% of the carbon per annum put into the atmosphere at present from the burning of fossil fuels (as discussed in Chapter 3).

To achieve a balanced situation, mature trees containing 4 million tonnes of carbon will need to be replaced by young trees each year. If, by a major investment, this felled timber could be transported to a number of wood—or, maybe more generally biomass—burning power generating stations (or alternatively plant to convert the wood to ethanol), this 4 million tonnes of carbon in the felled timber could be returned to the atmosphere each year but with the advantage of replacing an equivalent amount of fossil fuel. This would reduce the nett emission of CO_2 for a constant energy production, by about 2·5%.

Clearly, there is an ultimate limit to this option. If all the UK land mass was forested, it could still only recycle about one eighth of the present annual emission of CO_2 from fossil fuels.

It is the Working Party's view that there is merit in developing programmes along the above lines, even though the ultimate potential for reducing CO_2 emissions in the UK is not large, especially as the encouragement of forestry has other environmental as well as commercial benefits. It also has the merit of contributing to a worldwide development in the better use of woods and forests.

The rate of sequestration of carbon in UK forests is small compared with the carbon input to the atmosphere resulting from the tropical deforestation of between 1000 and 2000 billion tonnes/year, as discussed in the next chapter. None the less, it can be deduced from the CO_2 signal in the atmosphere that the productivity of the total boreal and temperate forest zone (of which the UK forests form a part) is having an increasing influence on the atmospheric CO_2 concentration, and the development of an expanding, efficient timber industry in the UK should be considered in this context.

REFERENCES

Department of the Environment (1989). *Global Climate Change*, HMSO 14 pp.
Forestry Commission (1988). *Forestry Facts and Figures 1987–1988*.
JARVIS, P. G., (1989). *Atmospheric carbon dioxide & forests*, Phil. Trans. R. Soc., London B 234 369–392.
McCULLOCH, A., (1990). Private communication.

Section 8

Scope for UK Assistance Outside the UK

It will be appreciated that, whatever the pressures to reduce the emission of CO_2 and the other greenhouse gases, there is bound to be an increase in emission over the next decade or so, at least, due to the increased use of fossil fuel worldwide. This will occur in order to meet the increasing demand for energy, particularly in the less developed countries.

There is considerable uncertainty as to the actual values to be put on this increase in energy production. It will depend on the state of the world economy and the success countries have in meeting their, often very ambitious objectives. Figure 8.1 shows one estimate by Yeager (1990), of the order of magnitude of the increased emission of CO_2 that it is anticipated the increase in fossil fuel consumption will produce, and it will be noted that an increased share of the total is to be expected from the use that developing countries will make, come what may, of their indigenous coal resources.

Another recent paper calculated that the emission of CO_2 globally could vary between about 9·5 and 11 billion tonnes carbon/year in the year 2000 and up to 14 to 24 billion tonnes carbon/year by the year 2050. The lower figures being produced by applying pressure through taxation on the use of fossil fuel to limit emission (Scott et al. 1989).

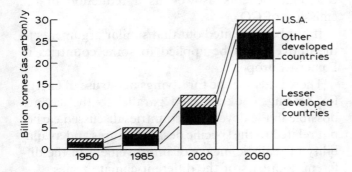

Fig. 8.1. Contribution to global man-made CO_2 emissions.

8.1 THE UK AND THE REST OF THE DEVELOPED WORLD

It is obvious from the figures given in Chapter 3 that the UK by itself is not going to make a significant reduction in the global emission of the greenhouse gases. Even if it was possible to reduce the emission of these gases from all the industrial and natural sources within these islands to zero, it would only reduce the total global emission by less than 3 per cent at the present time, a percentage that one must expect will decrease with time.

However, the UK will be expected to reduce its emissions as part of a joint effort with other developed countries, at least within the European Economic Community and hopefully as part of a concerted effort by all the developed countries including North America and the USSR.

If one considers CO_2 emissions alone, the percentage of the global emission emitted by the EEC countries, the USA & Canada, Japan, Australia and the USSR was, from the Table in Chapter 3, 67 per cent in 1987. The case for taking steps to reduce the emission of the greenhouse gases from the UK must therefore rest on supporting, possibly encouraging by example, and being part of a concerted effort by those countries who are already industrialised.

8.2 THE SITUATION IN DEVELOPING COUNTRIES

The situation is very different in those countries which have still some way to go to developing sufficiently enough to give their people a standard of living approaching that which the industrialised countries are used to, for example, in the UK. The situation is even more extreme in those countries where development into an industrialised society has scarcely begun but where the anticipation is that moves in this direction, in the least aimed at eliminating starvation and attaining some acceptable standard of living for the population as a whole, will be made over the next few decades.

It is clearly not possible to attempt to impose on these countries the same solutions to reduce the emission of the greenhouse gases which are appropriate in developed countries. In fact it can be assumed that, in these countries, priorities will be given to those moves aimed at improving the standard of living. They will insist on using their indigenous sources of energy and will be reluctant to divert any of their own resources away from the task of developing their own industrial—and hence energy—base.

At the same time, it is also already clear that the authorities in many of these developing countries are aware of the implications of some of the predicted effects of an increase in global temperature. Fear of the effect of a large rise in sea-level on a number of these developing countries has already been expressed. These countries are therefore, in general, ready to take part in any world-wide concerted efforts to reduce the greenhouse effect as long as it is within the framework of their other priorities and subject to help of both money and expertise from the developed countries. The UK will have its part to play in support for the developing countries and it is certain that placing finance and effort into some of the possibilities in these countries will produce a more significant reduction in the global emission of the greenhouse gases than equivalent expenditure within UK.

However, it is appreciated that there will not be clear cut alternatives such as 'is money to be put into measures in the UK or overseas'. As far as support to the developing countries is concerned, the priorities will be both to maximise the aid given and, also, to make sure such aid is used effectively to meet the real needs of the people while minimising all environmental ill-effects including those due to the emission of the greenhouse gases.

8.3 WHERE THE UK COULD HELP DEVELOPING COUNTRIES IN THE ENERGY SECTOR

Studies mentioned at the start of this chapter have shown that the rate of growth in energy production in the developing countries, predominantly from the use of indigenous fossil fuels, is outstripping that of the developed countries. The predictions are that this will increase over the next few decades as the developing countries strive to catch up with those already industrialised, and that the percentage of the global emissions of the greenhouse gases from the developing countries will rise from one-third in 1987 to more than half the total by early in the next century.

While it is hoped that maximum use will be made of available sources of renewable energy and biomass, much of the additional energy will be raised by burning indigenous coal, often of poor quality. Those countries fortunate to have their own sources of gas and oil will want to export as much of this as possible and it is difficult to think of arguments sufficiently powerful to stop them exploiting these energy reserves. There is clearly a great necessity for the UK, along with the rest of the developed world, to help ensure that this large increase in the use of fossil fuel is achieved with the minimum possible increase in the amount of CO_2 and other greenhouse gases emitted.

8.3.1 Use of Existing Plant

It is a common experience that much of the fuel burned in the under-developed countries of the world (as well as, it should be said, in many more developed countries) is used very inefficiently.

At the domestic end of the scale, there is the use of biomass, dung and fossil fuels and, especially these days, kerosene, in open fires. On the industrial scale, there are many plants, raising steam for electricity or process heating, which operate without working instruments and at efficiencies often half or less than that attainable if they were properly operated. The same statements can be made about many kilns and furnaces, especially once they have been installed a few years. Experience suggests that lack of spare parts and maintenance is the cause of this much reduced efficiency and that supplying these, coupled with training of both supervisory and operating staff, would produce both significant economic benefits as well as a reduction in the emission of CO_2.

It has been pointed out that similar arguments to the above could be applied to some countries in Eastern Europe.

There is also scope for savings in the use of energy for heating, cooking, etc., parallel to the savings possible in the developed countries discussed earlier but related to the specific design features and availability of realistic means of improvement in the different countries of the different climate zones.

Decentralised energy production (as opposed to the extended grid) has a bonus in encouraging the use of renewable energy techniques. The use of photovoltaic panels to supply a domestic electricity supply in remote areas can contribute significantly to the quality of life. Alternatively, a small wind turbine might be used. Both these options have a benefit locally out of all proportion to the effect on the overall energy demand of the developing country as a whole.

8.3.2 New Power and Steam Raising Plant and Other Furnaces

The large expansion in the use of energy envisaged in many developing countries will, however, only be realised if a major programme of building new power and steam raising plant and other furnaces is carried out. Contracting firms in the developed countries have and will be involved in the design and construction of these plants, many of which will be funded by international agencies such as the World Bank and the Asian Bank, who do have a real responsibility to see that their loans are spent prudently.

The requirements that these new plants be designed to minimise the emission of greenhouse gases will bring a new dimension to the responsibilities of the contracting companies and the funding agencies.

It will no doubt prove possible to persuade some of the developing countries to install plant using, in particular, natural gas rather than coal, especially where it is available within the country. There will, however, invariably be a cost to be picked up by the developed world to compensate for a loss of hard currency.

Where the only acceptable indigenous fuel is oil or, more generally, coal the best that can be done is to ensure that it is used as effectively as possible. This will be best achieved in medium sized, central power stations where a well trained local staff can be adequately supervised to ensure the efficient operation of reasonable but not unduly sophisticated plant. However, except maybe in a few special situations in countries that have already developed a high level of technical competence, it will not be wise to install the most energy-efficient systems such as coal gasification, combined cycle or pressurised fluid bed combustion plant until these are fully proven elsewhere. The developing countries should not be used as 'guinea pigs' for new developments.

There are however some advantages in encouraging the use of atmospheric (and in due course, pressurised) fluid bed combustion. This technology is effective in burning a wide range of fuels including fuels with variable composition.

8.4 AGRICULTURE IN THE DEVELOPING COUNTRIES

The use of energy in agriculture and the food industries in the UK has been discussed in Chapter 5 and this can be taken as typical for the developed countries. It was shown that appreciable energy is used in the manufacture of fertilisers and in the production of feed stuffs and in the processing of both crop and animal products in the food industry. However, it has to be remembered that agriculture is at present carried out very differently in the developing countries. Production is generally less intensive, uses fewer inputs of fertiliser, agrochemicals, machinery and fuel, and more intensive inputs of labour and animal power (itself indirectly derived from solar radiation).

It is to be anticipated that many of these developing countries will be under strong pressure to increase their output of food and it will be natural for them to look at and start to follow the same development path as that of the developed nations.

In so far as the use of fossil fuel, directly or indirectly, is a major contributor to the greenhouse effect such changes in agricultural practice on a large scale in the developing countries could contribute to a significant increase in the emission of the greenhouse gases. Devising feasible alternatives, acceptable and of benefit to the developing countries, should receive high priority. Alternatives to fertilisers would have to be based on biological nitrogen fixation, biological control of pests would have to be used instead of chemical sprays and, perhaps, increases in the use efficiency of animal power could reduce the need for fuel. Such lessons might well find applications in the developed world.

The question of the effect of crop production on the nett emission of the greenhouse gases has been discussed in Chapter 7. Only where longterm use is made of the vegetable by-products or biomass is there a nett reduction in the emissions or, at least, a delay in the release of gas back to the atmosphere. The use of timber and straw in building and manufacture is to be encouraged. The economics of using biomass as a fuel are adversely affected by its low density and calorific value. Collection is expensive and combustion equipment large. Similarly, while the conversion of biomass to liquid fuels is a well established technology, technical and, particularly, economic support will be necessary to persuade countries, particularly in the developing world, to indulge in such, generally major, projects.

In the longer term, there will no doubt be a pressing need for these uses of biomass to be developed, especially in the tropical zones of the world. As they generally require large and sophisticated plant their development should only be attempted if the right high degree of support is available.

The subject of forestry raises special issues which are discussed in the next section. Another crop which also raises particular problems is rice because of the release of methane from the bacteria that develop in the wet soil. Rice paddy fields may produce a quarter of all the methane emitted to the atmosphere but there is a surprising lack of reliable evidence on this. Estimates of the methane released by the oxygen intolerant bacteria in the water logged soils used for growing rice vary widely. Laboratory tests on soil samples from Japanese rice fields led to an estimate in the mid-1970's for the global annual emission of methane from this source of 280 million tonnes/y. Sporadic measurements on Californian rice fields suggested a figure of less than 60 million t/y methane emission. Continuous metering in Spain suggested similar or lower levels but work in Italy suggested values between 70 and 170 million t/y.

However, 90 per cent of the world's rice fields are in the Far East and there is remarkably little information on the amount of methane produced from these fields. What recent research has shown is that the temperature in the soil around the rice plant roots is important (Schutz et al. 1989): the warmer the soil the more methane that is produced. There is a further interesting suggestion that fertilisers have an important effect. Organic fertilisers increase the production of methane but chemical fertilisers can reduce methane formation. There is a good explanation for this in the case of sulphates as these would be expected to discourage the growth of methane producing bacteria.

This is a clear area where measurements are required across the main rice growing countries leading hopefully to the consideration of any way which might limit methane release without reducing the yield of rice. It is to be hoped that the appropriate UN Agency will initiate this work speedily.

Animals, especially ruminants, are also a major source of methane, especially as a result of the bacterial activity in the stomachs of the world's ever increasing herds of cattle. Quite a small proportion of the energy input to a ruminant is actually emitted

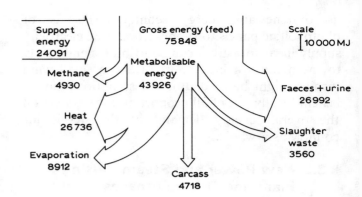

Fig. 8.2. The energy flow diagram for an extensive 24-month grass beet system (MJ per head).

as methane as can be seen from Fig. 8.2 (Spedding et al. 1983). However, the number of ruminant farm animals is very large and the total emission correspondingly high. The actual emission from an animal varies with its size, diet, rate of production and species but typical examples would be 400 litres per head per day from cattle and 50 l/head/day from sheep. If one takes the number of cattle and buffalo farmed in the world as 1.4×10^9 and the equivalent number of sheep as 1.6×10^9, this is an annual emission from cattle and buffalo of about 120 million t/y of methane and, from sheep, about 15 million t/y. This is higher than some other estimates, as discussed in Chapter 3, and illustrates the lack of data necessary to get an accurate understanding of the importance of this source of emission.

Other animals also produce methane, including camels, horses and pigs. However, the total number of camels, estimated at 19 million and horses and other equines, at 122 million is much less numerous than that for cattle and sheep and their individual output is lower. Pigs have a relatively low output of methane but the world population, estimated at 840 million, is quite high.

There is a minority group who would use this as an additional argument in favour of vegetarianism but the Working Party found it difficult to see what can realistically be done to reduce this source of emission or to stop it growing as living standards increase.

8.5 FORESTS, TROPICAL AND TEMPERATE

The role of trees in absorbing CO_2 as they grow and the release of this stored carbon back to the atmosphere when the tree is cut down and burned, or left to decay, has been discussed both in Chapter 2 and in Chapter 7, in relation to the UK. On the global scale, the recent large release of CO_2 to the atmosphere as the result of extensive cutting down and burning of large areas of forest, particularly in the tropics, where these forests are not being replaced by new plantings of trees, is a significant enough percentage of the net release of greenhouse gases to the atmosphere to merit separate attention.

Sustainable, mixed age, mixed species forest that is neither agrading nor degrading is, like agricultural crops, CO_2 neutral over 12 months. Such forest is to be found in a few extensive areas and in rather more small fragmented areas in the tropics, and in small fragmented areas of old growth temperate forest, particularly in North America and possibly also in Russia. Such forests are made up of a mosaic of small gaps of various ages in which sudden degradation has occurred and has been superseded by gradual agradation, the whole system being in CO_2 balance over areas of, say, 10×10 km and periods of, say, 10 years.

In contrast, rapid destruction over large areas by clearing of the forest and by fire results in the massive release of CO_2 to the atmosphere, usually in one or two large puffs over a short period of time, possibly followed by a slower release of CO_2 resulting from subsequent oxidation of organic matter in the soil. If afterwards agriculture is practised then the site will become CO_2 neutral again over a 12 month period but the standing crop of carbon that was present formerly in the trees and soil has been volatilised. The standing crop of climax tropical forest or old growth coniferous forest may be of the order of 500 tonne of dry matter per hectare and this can be expected to give rise to about 250 tonne of carbon that is transferred to the atmosphere once and for all.

There is very considerable uncertainty about the actual extent of deforestation that is taking place in the world. The best available data are based on the results of an FAO survey that depended on returns submitted by governments and was reported as long ago as 1980. For the tropics this amounted to about 11 million hectares per year between 1976 to 1980 (see Table 8.1) and represented annual clearance of 0·5 to 0·6 per cent of the area of forest per year.

Using the same data base, the net release of carbon to the atmosphere in 1980, resulting from deforestation and other changes in land use (Houghton et al. 1987), was estimated. The results are given in Table 8.2.

By far the largest proportion of this carbon input to the atmosphere in 1980 came from the clearance of forest in Brazil (18 per cent), Indonesia (11 per

Table 8.1 Deforestation in the Tropics, 1976–1980

	Closed Forests (Mha/y)	Open Woodlands (Mha/y)
Tropical America	4·12	1·27
Tropical Africa	1·33	2·34
Tropical Asia	1·82	0·19
Total	7·27	3·81

Table 8.2 Net release of carbon to the atmosphere, 1980

	Million Tonnes of Carbon Per Year
Temperate & boreal region	133
Tropical America	665
Tropical Africa	375
Tropical Asia	621
Total	1 794 ± 900

cent), Colombia (7 per cent), Ivory Coast (6 per cent), Thailand (5 per cent) and Laos (4 per cent).

Bearing in mind the considerable uncertainty attached to the above figures, the total amount represents almost a third of fossil fuel input to the atmosphere or a quarter of the total (i.e. $5·6 + 1·8 = 7·4$ billion tonnes of carbon per year).

In the tropics when agricultural land is abandoned, it is often the case that only a few years after the forest has been cleared, secondary forest comes in very fast. Growth of young forest, whether natural regeneration of tropical forest or temperate forest plantations, leads to accumulation of carbon in woody biomass again. Rates of accumulation can be rapid. In temperate regions, coniferous forest plantations can accumulate carbon at rates of up to 10 tonnes of carbon per hectare per year. Broadleaved forest would generally only grow at about one fifth of this rate, although poplars and eucalypts may grow up to twice as fast. Simple calculations suggest that a new large temperate forest of, say, twice the area of Europe could initially, i.e. over the next 40 years, absorb all the CO_2 currently being put into the atmosphere by deforestation in the tropics and utilisation of fossil fuels.

When the trees are eventually harvested, then the risk of oxidation increases enormously and there will be significant return of carbon to the atmosphere as CO_2. The rotation length for poplars may be only 4 years, for eucalypts 10 years, for conifers 60 years and for broadleaves 200 years. From the point of view of sequestering CO_2 from the atmosphere, slower growth over an extended period of time with a much reduced risk of reoxidation of the carbon, may be a better prospect.

The likelihood of reoxidation on the controlled harvesting of the trees depends very much on the utilisation of the woody product. Leaves, branches, bark and roots are usually left to reoxidise in situ over the next 10 years or more. The stem is usually manufactured. Clearly if the end products of manufacturing are disposable, such as tissue and newsprint, oxidation may occur very rapidly. If the timber is used for furniture or for building construction, oxidation is likely to be less rapid and particularly so for hardwoods in comparison to softwoods. From a CO_2 storage perspective, the only safe thing to do would be to put the stem wood into storage where oxidation would be unlikely (e.g. sunk in the oceans, or down coal mines). Oxidation of the products would not matter, however, if the global energy use system was a completely sustainable one based wholly on the utilisation of organic matter from crop residues and timber as a source of fuel: what should be oxidised today would be reassimilated tomorrow.

There are three major problems relating to the role of forests in the global CO_2 exchange:

(a) it seems that more carbon from forests is being oxidised to CO_2 today than is being assimilated,
(b) the large extra amounts of 'mined' carbon that are being returned to the atmosphere (5·6 billion tonnes per year), and
(c) an apparent discrepancy between the annual amount of the CO_2 evolved from forest destruction and the utilisation of fossil fuels and its subsequent fate in the atmosphere and oceans of the world.

It is to be hoped that international efforts may enable a balance to be achieved on (a) by both halting the destruction of forests in the tropics and stimulating the planting of new forests both in the tropics and temperate regions.

Planting new forests to take up an equivalent amount of CO_2 to that likely to be released over the lifetime of a new power station can contribute to the solution of (b). A specific example of this is the CARE Guatemala Agroforestry Project (Trexeler et al. 1989) where an area of 900 sq km of agroforestry is being developed to compensate for the CO_2 emission from a 180 MW coal-fired power station, i.e. over the 40-year life of the plant, 5 sq km of forest is required to compensate for the carbon emission in the production of 1 MW.

Finally, there is apparently a discrepancy between the total amount of CO_2 being put into the atmosphere annually of about 7·5 billion tonnes and the annual increase in the amount present in the atmosphere of about 3 billion tonnes. It has been supposed that the balance is absorbed into the oceans of the world but studies of biogeochemical cycling suggest that this is limited to about 2 billion tonnes per year. Consequently, the possibility remains that the forests of the world are a sink for CO_2 rather than a source as has been thought, on the basis of current information. Based on a very few measurements, it has been suggested, for example, that the Amazonian forest is not on average CO_2 neutral, as suggested above, but is a sink to the tune of about 1·2 billion tonnes of carbon per year.

In the UK the rate of accumulation of carbon in temperate forest plantations is currently doubling over the present 20 year period and it is likely that increase in area and growth of temperate forest is occurring throughout the Northern Hemisphere. One indication that this may be so is the increase in the peak-to-peak amplitude of the seasonal oscillation in the atmospheric CO_2 concentration at northern latitudes in the Northern Hemisphere. This signal may well indicate an increase in the biological activity of the temperate and boreal forest throughout the Northern Hemisphere, possibly as a result of CO_2 fertilisation itself, the rise in temperature experienced throughout the 1980's and deposition of nitrogen from the atmosphere. However this speculation, although based on plausible models, can in no way replace the urgent need for accurate information about the current areas of forest both in the tropics and in the temperate zones and their physiological activity in absorbing CO_2 on an annual basis.

There is, therefore, plenty of scope for UK experts both to increase our knowledge of the world's forest, and to apply this knowledge to the benefit of mankind.

REFERENCES

CANNELL, M. G. R., (1982). *World Forest Biomass and Primary Production Data*. Academic Press, London, pp 391.
DERRICK, A. (Pers. Comm.) Personal contribution to the Working Party 1990.
HOUGHTON, R. A., BOONE, R. D., FRUCI, J. R., HOBBIE, J. E., MELILLO, J. M., PALM, C. A., PETERSON, B. J., SHAVER, G. R., WOODWELL, G. M., MOORE, B., SKOLE, D. L. & MYERS, N., (1987). *The flux of carbon from terrestrial ecosystems to the atmosphere in 1980 due to changes*

in land use: geographic distribution of the global flux. Tellus, 39B 122–139.

JARVIS, P. G., (1989). *Atmospheric carbon dioxide and forests*. Phil. Trans. R. Soc. Lond., B 324, 369–392.

MARLAND, G., (1988). *The Prospect of Solving the CO_2 Problem through Global Reforestation*. United States Department of Energy, DOE/NBB-0082, pp 66.

SCHUTZ, H. et al., (1989). *A three year continuous record on methane emission rates from an Italian rice paddy*, J. Geophysical Research, 92, D13, 16405-16416.

SCOTT, M. J. et al., (1989). *Global Energy and the Greenhouse Issue*, 14th World Energy Conference, Montreal, Paper 2.1.1.

SPEDDING, C. R. W., THOMPSON, A. M. M. & JONES, M. R., (1983). *Energy and economics of intensive animal production*, Agro-Ecosystems, 8, 169–81.

TREXELER, M. C., FAETH, P. E. & KRANE, J. M., (1989). *Forestry and Response to Global Warming. An Analysis of the Guatemala Agroforestry Project*, World Resources Institute, Washington DC.

WHELDON, A. E., HANKINS, M., (1990). *Whats the use of twenty watts*, Energy World, 177, 13–14.

YEAGER, K. E., (1990). *Coal in the 21st Century*, Supplement to Energy World, 179, The Institute of Energy.

Section 9
Implications for Policy Formulation

9.1 INTRODUCTION

The Working Group's approach to the greenhouse issue was set out at the beginning of the report, and is repeated here to re-establish the basis on which conclusions are drawn from the preceding chapters.

It is not the Group's purpose to discuss the consequences of emissions of the greenhouse gases. Its purpose is to propose a range of measures that is likely to be most effective in bringing about a reduction in the rate of greenhouse gas emissions, should this become a requirement.

The Toronto conference in 1988, for example, called for a 20% reduction in global CO_2 emissions based on 1988 levels by the year 2005, and a 50% reduction from the same initial level by the year 2020. The Noordwijk Ministerial Conference on Atmospheric Pollution and Climatic Changes in the Hague in 1989 recognised in its declaration 'the need to stabilise . . . CO_2 emissions and emissions of other greenhouse gases not controlled by the Montreal Protocol'. It supported the opinion that emission stabilisation should be achieved at the latest by the year 2000, and endorsed the Toronto conference declaration.

In May 1990, the UK Government announced its objective to stabilise the emission of the greenhouse gases from the UK at their 1990 level by the year 2005.

9.2 REGULATION AND TAXATION

The effectiveness of measures can clearly be influenced by Government action on regulation and taxation. In the case of emissions of lead to the atmosphere, for example, the selective application of taxation has been highly effective in eliminating the use of two star petrol to free-up storage tanks for unleaded petrol, and in encouraging the purchase of unleaded rather than leaded petrol. Paradoxically, of course, the environmentally motivated move from leaded to unleaded petrol may be detrimental in greenhouse terms, because of the marginally poorer fuel consumption obtained.

Even though many of the measures highlighted later in this chapter as potentially effective in reducing the rate of greenhouse gas emissions would also save money if they were implemented, experience to date of the slow rate of penetration of these measures suggests to the Working Group that some added stimulus will be required if the potential of most of the measures is to be realised.

9.3 METHODOLOGY

In reviewing potential measures, the starting point is the assumption that the main greenhouse gases of significance in the UK at present are carbon dioxide, methane and CFCs. The Working Group discussed the various measures proposed in the preceding chapters, and their interactions in detail, and by consensus agreed a short-list of measures which in its view are likely to be most effective in the short-term. For the purposes here, the short-term was regarded as extending to say the year 2000. A second short-list of research topics for the longer term was also drawn up. In the Group's view these areas of research should be vigorously pursued. An attempt, which is not displayed here, was made to quantify the likely practical effect of the short-listed measures on emission rates of the greenhouse gases in order to test their effectiveness against the objectives set out in the Toronto and Hague declarations and in the UK Government's response. Some tentative conclusions are drawn. This is an area where further work is required. Significant points from the Working Group's discussion are recorded below.

9.4 CFCs

The signatories of the 1987 Montreal Protocol are now expected to agree at a meeting in London in

June to eliminate completely by the end of the century production of listed chlorofluorocarbons. It was reported in March that the signatories have also moved nearer to agreeing a financial package and a transfer of technology that would allow developing countries to switch to the production of less harmful substitutes. Despite this, the Watt Committee Working Group has included in the short-list measures to control emissions of CFCs because of their disproportionate impact on the greenhouse effect.

9.5 METHANE

Apart from emissions from ruminants, the major emission sources in the UK are from coal mining, refuse decay from landfill, and from oil and gas exploration, production and transmission.

Looking beyond the actions which are already in place or under review by the oil, gas and coal industries to minimise loss of methane to the atmosphere, the Group concluded that a substantial benefit could be obtained by the incineration of combustible, non-recyclable refuse and waste, in all possible cases with energy recovery. This measure is therefore included in the short-list.

9.6 ENERGY CONVERSION

The Group concluded that probably the greatest potential to reduce CO_2 emissions in the short-term in energy conversion lies in the installation of natural gas fired gas-turbines for industrial CHP and high-efficiency condensing combined-cycle electricity generation. These technological measures are therefore included in the short-list.

The Group considered that renewables should also have a place on the short-term short-list. From the cost-evidence currently available the major tidal barrage schemes seem unlikely to proceed if traditional financial criteria are applied. The renewable apparently with the best cost characteristics is wind power, which is therefore included in the short-list. The Group acknowledge that its contribution is likely to be small in the short term.

The future availability of natural gas and its price remain uncertain beyond the early part of the twenty-first century and hence the Group considers that it is essential for the longer term for research to continue into the clean and high efficiency use of coal in power generation, CO_2 removal from flue gas, and nuclear power. These technological options therefore feature on the second short-list for the longer term.

The development on a larger scale of wind power and of the other renewables, particularly solar energy, will no doubt provide increasingly useful sources of alternative energy but these will not, in the Working Party's view, be able to supply the bulk of the UK energy requirements in the forseeable future.

9.7 ENERGY SAVING

This is the area where substantial cost-savings remain to be obtained across all sectors: industry, public and commercial, and domestic. Some progress is usually assumed to be achieved in energy saving in projections of future energy demand, although on past performance, improvement will be slow in relation to the potential which exists.

In considering how progress could be accelerated, the Group concluded that the specification of higher standards (for example via the Building Regulations) both for new and for existing buildings could achieve substantial benefit. Public, commercial and domestic building insulation, draught proofing, heating and lighting, and electrical appliance efficiency were highlighted for particular attention.

9.8 TRANSPORTATION

The Group considered that the encouragement of smaller more efficient cars could have a substantial impact on CO_2 emissions. However, for encouragement to be effective, Government intervention in some form would be required.

For the longer term, research into new fuels and methods of transport should receive encouragement.

9.9 FORESTRY

Expansion of the forests to increase the fixed carbon reservoir was seen by the Group as having a small but significant role to play. (On the basis of the Group's crude quantification this measure may be several times more effective in the short-term than the promotion of wind power, for example.)

9.10 ASSISTANCE TO OTHER COUNTRIES

An example of the difficulties that will be experienced in implementing any greenhouse policy

Implications for policy formulation

globally is provided by noting what has happened since the breaching of the Berlin Wall. An area of East German industry which is predictably receiving prompt attention from the West is the motor industry, with the objective of tapping the vast potential market. At the same time, the opening up of Eastern Europe has shown the need for an enormous effort to improve environmental standards which can only be achieved with financial and technical help from the West.

In a more positive environmental vein, developed countries have a role to play in working with the less developed countries to formulate and implement global policies for forestry and agriculture, and to help improve the efficiency of energy conversion and utilisation. This collaboration need not, however, be exclusively one-way: the Philippines, for example, have implemented an effective transport policy which merits examination and may be more widely applicable.

9.11 THE SHORT-LISTS

Measures and research topics which the Group short-listed are shown in Tables 9.1 and 9.2.

9.12 DISCUSSION

Some argue that the UK's contribution to global greenhouse gas emissions is so small, that measures to reduce UK emissions by 20%, for example, are insignificant in a global context, and our limited resources would yield a better return deployed overseas in the less developed countries. There is no logic in this argument for measures which are sufficiently cost-effective in the UK such that their implementation would release rather than consume resources. Others argue that on account of the growth in energy demand in the developing countries the UK, if anything, requires to make a greater than proportional contribution to reductions sought in global emissions. Above all, the Working Group recognised that measures recommended for implementation in the less developed countries may lack credibility if they have not been implemented by the recommender.

Turning to the likely quantitative effect of implementation of the short-term measures listed in Table 9.1, the Group concluded on the basis of a rather crude quantification that against 'business as usual' projections of UK energy demand, which

Table 9.1 Short-list of short-term greenhouse measures

- High efficiency natural gas-fired gas-turbine combined cycle power generation plant.
- Natural gas-fired gas-turbine or gas engine combined heat and power in industry, the public and commercial sectors.
- The promotion of wind energy.
- Combustion of non-recyclable waste, refuse and biomass with energy recovery wherever possible.
- Energy saving in public and commercial buildings and the domestic sector. In particular by:
  - thermal insulation and draught-proofing
  - space-heating (efficiency, fuel type, CHP, controls)
  - more efficient electrical appliances
  - low energy lighting and lighting controls
- The promotion of smaller, more efficient cars
- Increasing the area of woodlands, especially combined with greater use of wood, including as a fuel.
- For listed CFCs:
  - adopt urgent measures to phase out manufacture world-wide
  - recover CFCs from plant and appliances so that they may be recycled or rendered harmless
  - substitute substances with a much smaller greenhouse potential than the listed CFCs.
- Work with the less developed countries:
  - to formulate and implement global policies for forestry and agriculture
  - to help improve the efficiency of energy conversion and utilisation

Table 9.2 Short-list of longer term research and development topics

- Clean and high efficiency use of coal in electricity generation
- CO_2 removal from flue gas
- Nuclear power
- New transport fuels and methods of transport

typically imply a 20% or so increase in CO_2 emissions between 1988 and 2005, the short-listed measures in Table 9.1 may well stabilise UK emissions at 1988–1990 levels in the short-term (i.e. by the year 2000). Further work is required, particularly to assess the degree of interaction between measures and the extent to which they are likely to be taken up, before a more definitive view can be given.

However, we need to be cautious over the efficacy of cost-saving energy efficiency measures actually to bring about a sustainable reduction in greenhouse gas emissions. The implementation of cost-effective energy measures may well initially bring about a reduction in greenhouse gas emissions, but as Dr Brookes, former Chief Economist to the UKAEA pointed out in the Jan/Feb 1990 issue of 'Energy World':

'Even if significant savings in fuels costs are achieved the money so saved must find an outlet and in industrial societies it is likely to be in purchases that consume energy in their production if not also in their use. We may conclude that although improvements in energy efficiency may bring benefits, reduction in total energy consumption is very unlikely to be among them.'

How do you discourage a householder who has saved money by improving insulation, draught-proofing and fitting low-energy light-bulbs from fulfilling his life-long ambition to take flying lessons? Indeed, should you try to discourage him?

Section 10

The View from Outside the UK

10.1 A COMMENTARY FROM DR G. R. WEBER, GESAMPTVERBAND DES DEUTSCHEN STEINKOHLEUBERGLAUS (GERMAN HARDCOAL MINING ORGANISATION)

This contribution addresses the following issues:

(1) Scientific assessment of the greenhouse issue
(2) World wide policy options to deal with global warming
(3) A German perspective

10.1.1 Scientific Assessment of the Greenhouse Issue

The current concern over the greenhouse effect, that is the possible additional greenhouse effect—and the resulting global warming—due to man's release of radiatively active trace gases into the atmosphere, is based upon computer model forecasts that project a 3°–4°C warming over the next decades. A warming of this magnitude is untested by human experience and could well have rather detrimental consequences on nature and human activities. The main ones are:

— Rising sea levels
— Shifting of climate zones
— Destruction of natural habitat

Those same models compute a temperature increase of close to 1°C as a result of the trace gas increase that has already occurred since the beginning of the industrial revolution. A critical assessment of temperature trends of the last 130 years, that includes the surface layer of the oceans as well as trends over the continents, shows that the actual, climatically relevant warming could only have been in the neighbourhood of 0·3°C and therefore falls drastically short of the model predicted temperature rise.

Moreover, the spatial and temporal pattern of that warming cannot be explained by any of the currently available climate models.

Furthermore, it is not justifiable to ascribe that observed smaller warming entirely to the greenhouse effect, because of the presence of natural factors in the climate system, which do also point to a warming during the last 100 years. And finally, it is well known from climate history, that temperatures fell prior to the rise beginning in the middle of the 19th century due to natural causes. Consequently, a subsequent rise cannot automatically be viewed as a bona fide proof of the greenhouse effect.

10.1.2 World Wide Policy Options to Deal with Global Warming

This is the backdrop against which model predictions of a further large rise of global temperatures and the proposed measures to ward off that rise should be viewed. We are within a realm of uncertainty: Temperatures may rise and then again they may not rise—or not nearly as much as predicted by the models, which at this point appears to be the most likely outcome.

Yet mankind is obliged to act within this realm of uncertainty.

Applying a measure of caution to both our global environment and to unproven model forecasts, it is our task to find the right kind of response. Wedged in between the risks of costly overreaction on the one side and equally costly underreaction on the other, we must ensure that any preventive and abative strategy will not place a greater burden on mankind than the envisioned climate changes themselves and fails to provide sufficient energy to an energy hungry world. It is wise then to adopt those measures first which help the environment and which also benefit us in other ways. First and foremost we must realise that global warming is

—unlike traditional air-pollution problems—an issue of global scope. No national government could solve that problem in a go-it-alone fashion. International, world wide co-operation is necessary if any success in reducing trace gas emissions and curtailing the greenhouse effect is to be achieved. When reducing energy related CO_2-emissions, even the EEC's role, which accounts for approx. 14% of total CO_2 emissions, is severely limited.

Therefore, unless there is an internationally negotiated and agreed upon, equitable and verifiable framework of reducing CO_2 emissions—if one wants to focus on CO_2 alone—chances for globally reducing—or at least slowing down—an atmospheric CO_2 build-up do not look very promising. Focussing on the available options to actually reduce CO_2 emissions—once a decision to do so has been made—employing technological means to increase energy efficiency seem to carry the greatest promise.

The last 15 years have given us a valuable historic precedent of how industrial societies became more energy efficient (emit less CO_2 per unit of GDP) under the threat of a cut-off of energy supplies—or at least under the shock of a steep price increase. It should be remembered though that this steep price increase brought with it deleterious consequences for the general economy, and it may therefore not be very wise to subject the economy once again to the shock treatment of an artificial 'energy crisis' via a strangulating energy tax.

Several studies indicate that, from an investment point of view, the first Pound Sterling (Dollar, Mark, Franc, Yen) invested in energy efficient technology is the one most efficiently invested. As a result, after 15 years of investing in energy efficient technology, there may already be countries—or industries—where an additional increase in energy efficiency (and CO_2 reduction) can only be achieved —though technologically possible—at prohibitive cost.

Assuming that the available capital stock is limited, it is much wiser then to invest the money there, where maximum reduction per dollar invested can be achieved. This is the essence of the clearing house concept. Therefore, and returning to the beginning, the international response to global warming should be a staged one, where the first stage would be the application of those measures which are cost neutral or even cost negative. In subsequent stages the extent of those measures should be subjected to a regular review process and altered according to what the ongoing scientific assessment of the greenhouse issue mandates: a tightening of reduction measures, if the scientific evidence points to a larger warming than hitherto expected and a relaxation if the evidence points to a smaller warming. The setting of specific reduction targets does not presently appear to be justified given the current inconclusive scientific evidence on global warming.

10.1.3 A German Perspective

Germany accounts for less than 4 % of global energy related CO_2 emissions and could consequently not in any way significantly alter the global CO_2 balance by itself. Only within internationally binding agreements could Germany play a role.

To probe into the greenhouse effect and into options averting it, a parliamentary commission has been set up. Assuming that we do indeed face the large warming climate models expect, a range of measures Germany alone could adopt has been analysed. As a result, it has been suggested, that expanded reliance on nuclear power may significantly curtail CO_2 emissions but will certainly not solve the problem. Likewise, extensive—partially cost-prohibitive—measures to increase energy efficiency could also achieve the desired result. Expanded reliance on renewable energy was deemed unrealistic given the magnitude of Germany's energy needs and the country's unfavourable geographic location with respect to using solar energy. Most forms of renewable energy as well as some of the suggested technological solutions become economically competitive only at much higher energy prices.

However, the conclusion commonly drawn from this, namely 'energy must be more expensive' so that renewables and expensive technologies are economically competitive, must be viewed with utmost caution. It certainly cannot be in the best interest of any industrialised society, including Germany, to relentlessly increase the price of energy. Yet, for a variety of reasons, it is in their best interest to use energy rather miserly. Therefore, avenues have to be explored, which keep both the price and the consumption of energy and, in the case of fossil energy, the attendant emission of gaseous substances, low.

Energy experts in Germany think that the efficiency of current fossil fuelled power plants can only be increased by a small amount, but point out that the next generation of fossil power plants, which will be available after the turn of the century will be considerably more energy efficient.

From an investment point of view though, it may presently be better to apply the available capital stock to the rather inefficient energy systems in Eastern Europe. Amid the ongoing political change this may indeed be the better way, achieving several objectives at once:

— Giving economic aid may help to politically stabilize the region
— Clean up the local environment from traditional air pollutants (suspended particulate matter, SO_2, etc.)
— Reduce CO_2 emissions and help stave off the greenhouse effect.

Expanded use of nuclear energy in the electricity generating sector, which currently makes up about one third of Germany's CO_2 emissions, may appear unpalatable to the general public. Insofar it is unlikely that Germany's national energy policy, which rests on the principles of security of supply, economical competitiveness and environmental compatibility, will be re-written on the basis of the CO_2 issue alone. Germany has to continue to rely on a diversified mix of energy sources. Each of those sources has endemic problems which have to be solved for each type of energy individually.

As far as fossil fuels are concerned, increased efficiency can and will be achieved by continually advancing technology. An extra bonus can be expected from the transfer of modern, western technology to Eastern Europe and East Germany.

10.2 A COMMENTARY FROM M. NOMINE, CENTRE D'ETUDES ET DE RECHERCHES DE CHARBONNAGES DE FRANCE

The comments expressed here must not be considered as the official position of France in the debate. They reflect only the personal views of M. Nomine on the subject.

10.2.1 Emission of Greenhouse Gases

10.2.1.1 CO_2

The situation in France is somewhat different from the one in the United Kingdom. For an almost equivalent overall global energy consumption, CO_2 emissions expressed as tonnes carbon is only 100 MT in France compared to 160 MT in the United Kingdom. This leads to a low specific emission of 1·8 tonne C per capita which compares to higher figures in the USA (5), in Germany (3) and in the United Kingdom (2·5).

Table 10.1 Evolution of electricity production in France

Year	CO_2 producing electricity generation as a % of the total production
1970	54·6
1972	62
1974	60·3
1976	66·2
1978	55·6
1980	48·2
1982	34·2
1984	19·9
1986	11·4
1988	8·5

The main reason is obviously the extent of the Nuclear Programme for electricity production as illustrated in Table 10.1, where it can be seen that CO_2 producing electricity generation is less than 10% of the total in 1988. For the same generation structure as in the United Kingdom, France should produce about 50 MT carbon more than the actual figure. Another reason is the vigorous policy of energy conservation that has been undertaken since the first oil price shock, with the creation of the French Agency for Energy Conservation. This policy made energy intensity decrease by 20% between 1973 and 1988, of which 17% took place between 1973–1980, which means that conservation efforts have had less impact in recent years.

It is interesting also to compare the situation in

Table 10.2 Primary energy consumption by source (as Tec % of total)

Source	France (Tec % of total)	UK (Tec % of total)
Coal	9%	32·6%
Oil	43·5%	34·1%
Gas	12·1%	24·7%
Primary Electricity	35·4%	–
Others	–	8·5%

Table 10.3 Carbon emissions by sector and fuel type as MTc/year

Sector	Coal	Oil	Gas	Electricity	Others
Industry + Iron and Steel	9·2	7·2	5	2·3	ε
Domestic + Tertiary	3·1	13·9	7·6	3·2	1·6
Transport	0	31·1	0	0·13	0

France and in the United Kingdom regarding primary energy consumption (Table 10.2) and the carbon emission by sector and fuel type (Table 10.3). One can notice that the fuel mix is quite different, the share of oil being comparatively higher in France. The same applies to carbon emissions by sector and fuel type: the most important emitters are clearly the domestic and the transport sectors. This may have a significant impact on emission reduction policies.

10.2.1.2 Other greenhouse gases

Reliable emission estimates are not readily available today. Work is going on, under the responsibility of the Ministry of Environment, to establish emission factors relative to CH_4 and N_2O on the basis of published data.

As the emissions are mainly linked with agricultural practices (CH_4, N_2O) and standard of living (CFC, NO_x, etc.) there should not be major differences in global emissions between France and the United Kingdom. This means that the influence of non CO_2 greenhouse gases is comparatively higher in France than in the United Kingdom.

10.2.2 Implications for Policy

Although responsible for only a small percentage of worldwide greenhouse gases emissions, France will obviously be involved in negotiations on an international global agreement on emissions limitation. The Ministry of Environment has appointed two working groups in charge of assisting in policy formulation.

One is composed of members of the Academy of Sciences. Its task is to establish whether the problem of global warming is serious enough to justify the implementation of preventive measures without any delay. The other is composed of officials from different Ministries (Environment, Industry, Trade, Agriculture, Public Health, etc.) with the assistance of technical experts. This group is in charge of assessing the impact on the economy of the practical solutions that can be considered and on the means by which the objectives can be achieved (carbon taxes, fuel use limitation, tradeable permits, etc.).

A report is expected to be issued in 1990 and should form the basis for the French position in the forthcoming international negotiations. Regarding the technological responses to the greenhouse effect, France is considering all the options listed in Chapter 9 of this report.

10.2.2.1 Methane

Emissions from coal mining, oil and gas production are inherently low due to the low level of these activities in France. Losses of methane during transportation and distribution are strictly controlled by the distribution company GdF since they represent significant loss of income. Methane recovery from closed underground mines seems economically attractive at the present price of gas, at least in certain locations. Financial incentives could make it even more attractive.

As regards emissions by refuse decay from landfills, the situation is quite confusing. It is more and more difficult to establish new landfill sites due to the strong opposition of local populations, but it is also and perhaps even more difficult to install waste incinerators because of strong concerns about suspected air pollutants emissions (dioxines, PAH, heavy metals, etc.).

10.2.2.2 Energy conversion

The remaining potential for conservation is not as great as in the United Kingdom.

Conversion of electricity generation capacity to non CO_2 producing technologies has been already completed. It seems quite impossible to decrease further the proportion of coal and oil generated electricity in France. On the other hand, it is possible to contribute to conversion in other European countries through nuclear electricity exports since the existing capacity is not yet saturated (in normal weather conditions). Some commentators are prepared to consider the installation of additional nuclear capacities to increase exports further. Whether this will be acceptable by the population is another story.

For the near future, projections on electricity demand show that some fossil fueled capacity should be necessary for semi-base load (less than 2500 hr/year) by the turn of the century. One of the options is to build large circulating fluidized bed boilers that should meet the requirements of flexibility and conformity to EEC environmental standards. Attention has to be paid however to the fairly large emissions of N_2O by these boilers as compared to conventional PF firing.

Other options considered are repowering of fuel oil fired boilers with additional gas-turbine capacities. Integration gasification combined cycles are not considered at all for the time being.

Conversion of electricity generation to renewable should be only marginal. Hydraulic equipment can be considered as completed. A project like the tidal barrage of the Mont St Michel's Bay is unlikely to get approval because of the strong opposition of environmentalists.

There exist limited but significant possibilities in combined heat and power generation through municipal waste burning but, as mentioned above, the problem of public concern about potentially toxic air pollutant emissions has to be solved. Another possibility is the development of waste methanisation processes like the Valorga process to produce methane that can be burnt cleanly.

10.2.2.3 Energy savings

As mentioned previously, much has been done in this sector since the oil crisis. This means that the most cost-effective, low payback investments for energy conservation have already been achieved. At the present relative price of energies, there is not a strong incentive to invest more although there are still large technical possibilities. Some kind of incentive or price alteration would be necessary to make energy saving more attractive.

10.2.2.4 Transportation

From Table 10.3 it can be seen that the relative weight of the transportation sector in national CO_2 emissions is fairly high. Moreover, fuel consumption has been continuously growing although the conventional specific fuel consumption of personal vehicles has decreased by 25% since 1975. Undoubtedly one should pay special attention to this sector, which is not the easiest to manage, particularly as CO_2 reduction possibilities rely almost exclusively, at least for the short to medium term, on a change in the behaviour of people. Although difficult, any effort towards fuel consumption reduction or at least stabilisation must be encouraged.

10.2.2.5 Forestry

There are large possibilities in France to expand forests to increase the carbon reservoir since part of the cultivated land is supposed to go back to fallow. A positive approach to greenhouse gas emissions reduction could also be to improve control measures to prevent forest fires.

10.2.2.6 Assistance to other countries

There is no doubt that France will be strongly involved in actions aimed at improving fuel use efficiency in less developed countries, without impairing their attempts to develop. It is recognised that the participation of less developed countries is crucial for the implementation of a global agreement on greenhouse gases.

10.2.3 Conclusions

Although its technological responses to the greenhouse effect can be different from those of other countries, it is quite certain that France will be involved in the global efforts to decrease, or at least stabilise emissions. France will also play a role in the efforts to understand the phenomenon better and to predict the effects.

10.3 A COMMENTARY FROM DR. ANTONIO DURAN, EMPRESA NACIONAL DE ELECTRICIDAD, S.A. (ENDESA), SPAIN

10.3.1 Main Environmental Concerns in Spain

In Spain, as in most industrial countries, there are serious problems in connection with the quality of the environment. Concern about industrial pollution is considerable. For instance, the problems involved in the environmental management of river and coastal waters, of waste disposal sites, and in the use of low quality domestic coals are well known. However, the most pressing environmental problems in Spain today are increased desertification of the land and growing deforestation owing to forest fires.

An unfavourable climate, with low rainfall and high evaporation rates in broad areas of the country, a severe land erosion, loss of agricultural soils and a number of other negative factors have brought about an increase in the number and size of desert areas. Forest fires contribute to worsen this situation by destroying unique environments. According to ICONA (Spanish Institute for Nature Preservation) data, during the 1978–1987 decade an average of 8100 fires occurred every year. The average land area affected by fire was 260 000 hectares/year, representing about 0·5% of Spanish territory. An average of 105 000 ha/year of burnt areas corresponds to woodlands. The real impact of these facts must be studied in more detail in the future in order to evaluate their actual contribution to the greenhouse problem.

As previously stated, this does not in any way mean the absence of other environmental problems whether from industrial or residential sources, but on account of specific Spanish conditions graphy location, climate, demography, etc.), their effects are generally limited locally and scarcely severe.

10.3.2 Greenhouse Gas Emissions in Spain

To date, there is not a detailed inventory of Spanish greenhouse gas emissions. The large number of sources—both natural and man-made—and the variety of greenhouse gases make this evaluation difficult. The above notwithstanding, some studies have been undertaken to fill this gap.

The estimation of carbon dioxide emissions caused by the use of fossil fuels and other industrial activities is an easier task, even if only approximate

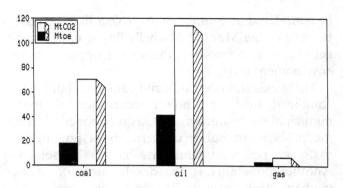

Fig. 10.3. CO_2 emissions and primary energy uses in Spain (1987).

accuracy is achieved. According to data included in the 1987 Spanish Energy Balance and from cement manufacturers' figures, CO_2 emissions are estimated at about 204 million metric tons (55·6 Mt of carbon). Out of this total, 193 Mt of CO_2 comes from the use of fossil fuels. Carbon dioxide emissions from these sources amounts to 5·3 t per capita (1·4 t of carbon). Figure 10.1 shows CO_2 emissions broken down by sector; Fig. 10.2 represents emissions by type and usage of fossil fuels; and Fig. 10.3 represents emissions by primary energy uses.

It can be observed that Spanish CO_2 emissions, when compared with global emissions, contribute about 1% to the total; Spain takes 19th place among the world countries (Fig. 10.4).

It is pertinent at this point to briefly examine the situation of power generation in Spain. At the end of 1987, the installed electricity productive capacity

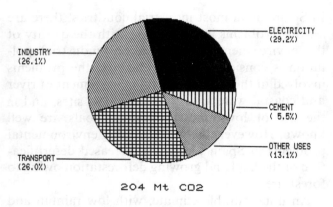

Fig. 10.1. CO_2 emissions in Spain (1987).

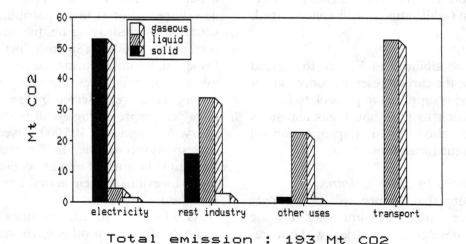

Fig. 10.2. CO_2 emissions in Spain from fossil fuels (1987).

The view from outside the UK

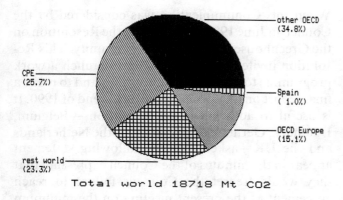

Fig. 10.4. Total world CO_2 emissions from fossil fuels (1987).

was about 41 000 MWe, with 46·7% of that total corresponding to conventional coal or oil-fired power plants (Fig. 10.5).

From the standpoint of electricity production, 97% was contributed by coal-burning and nuclear plants (Fig. 10.6).

Carbon dioxide emissions from the electric sector for 1987 were 59 Mt and over 89% of this originated from the combustion of the different types of coal

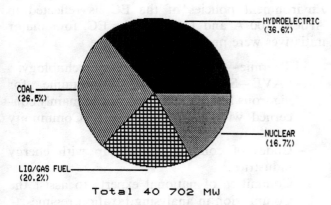

Fig. 10.5. Installed electrical power in Spain (1987).

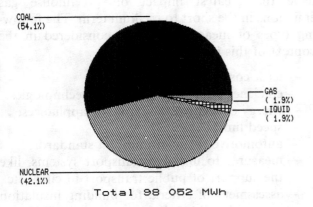

Fig. 10.6. Electricity generation by fuel type (1987).

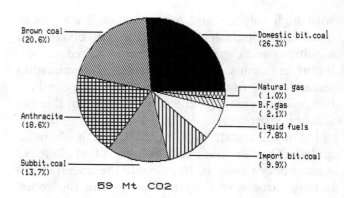

Fig. 10.7. CO_2 emissions from power generation in Spain (1987).

(Fig. 10.7 shows the share of the various fossil fuels). This means that the electric sector contributes 29% to total CO_2 emissions in Spain (about 0·3% of the total global CO_2 emissions). The specific emissions (average) from fossil fuel burning power plants amount to 0·96 kg CO_2/kWh.

At this time, there are not yet any evaluations of other greenhouse gas emissions in Spain.

10.3.3 Projections and Lines of Action on the Greenhouse Issue

The increasing concern about the greenhouse effect has been taken into account by the Spanish authorities and policy makers and will be considered in the 1990 National Energy Plan, presently under study. This Plan, to be approved in the last months of this year, will establish the basis and guidelines for energy policies and strategies for the next years. As this Plan is now defining the future structure of the power generation system, it is not possible at this time to evaluate its effects or implications on the limitation of greenhouse emissions. In any case, it is clear that power generation using coal will keep a preferred position as several plants will be built or repowered, and these plants will burn either imported or domestic coal. An increase in the use of natural gas is possible but, at this time, the share of nuclear power plants in the whole electricity generation scheme remains unknown.

Today, the energy sector, and particularly power generation using fossil fuels, faces serious environmental problems. As a main concern, the EC Directive on the limitation of atmospheric emissions from large combustion plants raised the urgent necessity to adopt technical and economical measures to reduce SO_2 and NO_x emissions from existing and new power plants. This is an especially difficult task in respect of the abundant low quality domestic coal with high sulphur and ash contents. As a consequence, the Spanish electric sector is involved in an ambitious and expensive programme to reduce pollutant emissions by using advanced combustion technologies, for example, fluidised bed combustion, coal cleaning systems, fuel blending or flue gas scrubbing processes.

In this context, the limited Spanish share in global CO_2 emissions could be seen at first glance as a secondary problem. But despite the uncertainties in many aspects of the greenhouse issue, the implementation of preventive measures has to be considered as proper action if included in an international framework.

Several measures included in the National Energy Plan—and also discussed in the present Report—will obviously have a positive effect in the reduction of CO_2 emissions. These include:

— improved energy efficiency;
— use of co-generation and combined cycles;
— fuel usage strategies;
— use of renewable or alternative energies;
— implementation of R & D programmes, etc.

but, as mentioned earlier, a more accurate knowledge of future lines of action is not possible at this time.

Moreover, forest fire prevention—certainly a very difficult task in Spain—and the reforestation policies carried out by the Spanish Government and local authorities will have a highly favourable effect in reducing the greenhouse effect.

It may be said as a final comment that, in spite of the natural differences in the environmental situations of the United Kingdom and Spain, the study undertaken by the Watt Committee on Energy will be a very useful guide to identify and evaluate more accurately the different sources of greenhouse gases as well as to analyse the wide field of possible actions to cope with their reduction. In this respect, the discussed methodologies can be used to approach the Spanish case.

10.4 A COMMENTARY FROM THE DIRECTORATE FOR ENVIRONMENT, NUCLEAR SAFETY & CIVIL PROTECTION OF THE COMMISSION OF THE EUROPEAN COMMUNITIES

Energy has a major input into the emissions of the greenhouse gases. This was recognised by the EC Commission in their first Communication on the greenhouse effect to the Council in November 1988. When this Communication was considered by the Council in June 1989 they adopted the Resolution on the Greenhouse Issue and the Community. This Resolution invited the Commission to launch a work programme to evaluate policy options and to make a final report on that programme by the end of 1990. It is useful to add that seven delegations—Belgium, Denmark, Germany, Luxembourg, the Netherlands and the UK—asked that the following statement appear in the minutes of the Council expressing that they would 'regret the Councils failure to reach agreement at the present meeting on the minimum target of stabilisation of emissions contributing to the greenhouse effect, and in particular CO_2'.

In adopting the 'small cars' directive in June 1989 (Dir 89/458/EEC) the Council requested the Commission to make proposals to limit the CO_2 emissions from vehicles by way of Article 6 of that directive.

The next milestone on the Community's approach to the problems posed by the greenhouse effect was the Commission's Communication to the Council—COM 369(89) Energy and the Environment. In this Communication Treaty, which seeks to form the basis of the integrated Energy and Environment policies of the EC as reflected in Articles 100 A and 130 R of the EC, four major initiatives were proposed. These were:

— Thermie—promotion of energy technology.
— SAVE—Specific Action Programme for vigorous Energy Efficiency will be mainly concerned with legal measures at Community level.
— Codes of conduct—drawn up with energy industries.
— Committee of national experts to assist the Commission in analysing taxation regimes.

It is the second initiative—SAVE— that should have the greatest impact on greenhouse gas emissions in the short to medium term. The following types of measures may be considered in the context of this initiative:

— least cost planning,
— efficiency standards for energy technologies,
— energy efficiency standards for appliances,
— speed limits,
— automotive fuel consumption standards,
— measures to improve transport systems, like the support of public transport in cities etc.,
— assessment of criteria for building insulation standards taking into account the different climatic conditions in the Community,

— elimination of legal and economic barriers to facilitate/increase sales of power/heat to energy distributors and to end users,
— minimisation of methane leakages from natural gas distribution systems.

The Commission will propose a coherent global programme on energy efficiency improvements and energy conservation to be presented as a follow up of this Communication by early 1990. This special SAVE programme will cover a multiannual period and will have to address priority areas of concern, as well as actions that can quickly yield results. Measures need to be identified which represent the most efficient way of action at Community level and which give the highest value added results.

The Commission has recently sent another Communication to the Council on 'Community policy targets on the greenhouse issue'. In that Communication, reference is made to the fact that, given the hesitations of many industrialised countries, the Community should take the lead along the lines of the Resolution. To achieve this objective, it is necessary to expand, develop and strengthen some of that Resolution's statements. In particular, the Commission underlined the following:

(a) Clearly, the long term objective of an overall greenhouse policy should be to stabilise atmospheric concentrations of greenhouse gases at levels compatible with acceptable climate conditions on earth. It is not yet possible to fix such levels in a scientifically sound way. However, the strategy of just waiting for a global long term response policy, which could only be established based on a better long term understanding of the complex science involved and on a full assessment of possible courses of action, is not acceptable and must be complemented by a pragmatic and adaptive approach. *Immediate measures should be taken now on obvious priority items*. Main candidates for such action (apart from CFC's which are dealt with under the Montreal Protocol) are CO_2 emissions in industrialised countries and protection of tropical forests.

(b) Therefore, for a start, the Community should underline the urgent need for *a clear commitment by industrialised countries to stabilise their CO_2 emissions by the year 2000*. Such stabilisation should be, in principle, at the present emission level. However, those industrialised countries which, as yet, have low energy requirements, and which can reasonably be expected to grow in step with their development, might have targets which accommodate that development.

(c) Moreover, *significant CO_2 emission reductions at the horizon 2010 by industrialised countries should be considered*. To this aim, policy targets should be defined in the light of results from the IPCC assessment presently in progress and of results of Commission evaluations.

(d) In the light of the above statements, *urgent action to limit CO_2 emissions is needed*. To this end, based on proposals by the Commission, the Council should in particular consider, before the end of 1990, measures to limit CO_2 emissions from vehicles, by improving their energy efficiency, as announced in Directive 89/458/EEC (O.J. L226 of 3.8.89), Art. 6. Moreover, in the frame of the SAVE (Vigorous Action to Save Energy) programme, announced by the Commission in its 1990 work programme, immediate, concrete and vigorous action will be launched in particular priority sectors where a significant increase of energy consumption could take place in the near future.

(e) In general, the overall policy to be established should ensure, by an optimal use of both economic instruments and standards, that, over time, external costs to the environment are fully taken into account.

(f) The Community and Member States should promote the adoption of an appropriate international agreement consistent with the objective expressed in (b) to (e) above, in the context of a global climate convention to be considered by the 1992 United Nations Conference on Environment and Development.

(g) The Community and Member States should take appropriate initiatives to promote coordinated and cooperative action with third countries. In particular Eastern European and developing countries. Such action should in particular be related to transfer of technology and to the implementation of energy policies aimed at greenhouse gases emission reductions.

(h) The Community should underline *the need to halt deforestation by the year 2000* and to reverse the present trend by that date and should confirm its willingness, with the means at its disposal, to fully contribute to a comprehensive, coordinated action on tropical forests. In conformity with the conclusions of the Noorkwijk conference, a world growth of 12 million hectares a year in the beginning of the next century should be considered as a provisional policy target in this field.

Studies to be launched in 1990 will include:

1. Policy measures aiming to reduce CO_2 emissions from the transport sector.
2. Policy measures aimed at reducing CO_2 emissions in the electricity sector.
3. Policy measures aimed at reducing CO_2 emissions from the most relevant industrial sources.
4. Policy measures aimed at reducing CO_2 emissions in the commercial/domestic sector.
5. Measures to share the effort of limiting greenhouse gases among countries internationally and within the EC.
6. Use of economic instruments to limit CO_2 emissions.
7. Assessment and analysis of CO_2 emissions by sector in the EC.

Current Community's R&D programmes pertaining to global change will concentrate on the following areas:

Environment
— EPOCH (European Programme on Climatology and Natural Hazards)
— STEP (Science and Technology for Environmental Protection)
— MAST (Marine Science and Technology)

Energy
— JOULE (Non-nuclear energy, including energy and environment modelling)
— Nuclear Safety

Economics
— SPES (Stimulation Programme in Economics)
— FAST (Forecast and Assessment for Science and Technology)
— SAST (Strategy Analysis for Science and Technology)

The Joint Research Centre develops its own research on the environment, on global changes and on remote sensing.

These activities will be expanded under the 1990–94 Framework Programme. European Community's funding for research related to global change issues, for the next three years, exceeds 300 million ECU (about 360 million US$) which represents, through cost-sharing arrangements, a volume of research over 600 million ECU (about 720 million US$).

In the non-nuclear energy field, R&D is supplemented by demonstration and technology promotion actions (THERMIE programme) which has already been referred to.

Section 11

Closing Remarks

This study has proposed a range of measures that the UK might adopt to bring about a reduction in the rate of greenhouse gas emissions which could at least meet the UK Government's present target to stabilise the rate of emissions at the present (1990) levels by the year 2005. It has further proposed research and development to be started now that might lead to the acceptance of other technological changes which could enable these levels of emission to be held or, possibly, reduced up to at least the year 2020—that is over the next thirty years from the date of this Report.

The Working Party was criticised by a few participants in the Consultative Conference for not proposing more radical actions (Mellanby, Pers. Comm.). It remains the Working Party's view that it will be difficult enough to achieve the reduction proposed above within the proposed time scale, especially in the absence of any decisive evidence that the build up of the greenhouse gases is causing global warming and the realisation that, for these limitations in emissions to have any influence on global warming, they must be applied world wide. There is bound to be a delay while international agreements are negotiated and, it is to be hoped, satisfactorily concluded. In the meantime, unilateral action by the UK, while no doubt desirable as an example to others, must be tempered by the realisation that such action in isolation will solve nothing and a realistic assessment of the economic consequences of the measures promoted.

The options proposed by those seeking a more dramatic and radical approach are all included in this Report. The difference is in the speed and completeness with which it is proposed they should be implemented. To take the most common radical approach, the abolition of the use of fossil fuels world wide or, in some proposals, by the developed countries only, would cause a wide and unacceptable deterioration in living standards if the energy that would be produced by fossil fuels were not replaced by alternative sources of energy (after allowing for the savings in energy to be achieved by energy use efficiency and similar measures). The review, given in this Report, of alternative sources of energy suggests that those presently technically available will be insufficient to meet this demand with the possible exception of nuclear power. However, assuming that alternative processes are available and, furthermore, that the enormous capital expenditure is immediately available for the plant that will be required, it will still take most of the thirty years up to 2020—if not longer—before such a major rebuilding of the energy supply industry, either in the UK or worldwide, can be achieved. Any prediction of what the world will be like thirty years and more from now is bound to be wrong.

Ideas for reducing global warming or the concentration of the greenhouse gases in the atmosphere which, at present seem unrealistic may by then have been shown to be feasible. Encouraging the growth of algae in the oceans to absorb CO_2 and possibly harvesting the algae as fuel, is one such proposal. Other actions, such as painting the world's deserts white or filling the oceans with white polystyrene balls to reflect the sun's heat, may appear slightly more realisable at the present time than putting in space a giant aluminium umbrella twenty times the size of Britain. There are other ideas, and there will no doubt be more, some of which will prove valuable and viable if the need is great enough.

The demand for energy will have grown as people, especially in the undeveloped and the developing countries of the world, increase their living standards, and also because of the increasing world population. It is clear that this increasing population—assuming nothing happens to limit its apparently inexorable rise—will raise in time a whole host of intractable problems of which the increased emission of greenhouse gases will be only one. What, if anything, can be done about this question is beyond the scope of the present report.

This increased demand for energy will hasten the day when, as is inevitable, the world reserves of oil and natural gas will be seen to be running out. This will, as far as our present knowledge of the available technology tells us, leave coal and nuclear power as the only two energy sources likely to be able to supply the bulk of the world demand.

Representatives of the nuclear industry claim with some justification that nuclear power is already developed and available (Donaldson, Pers. Comm.). It is also not a significant emitter of carbon dioxide. It is a fact, however, that the growth of the nuclear industry in the UK and worldwide is generally dormant. The Working Party recommends in a short-list of longer term Research and Development Topics (Table 9.2, Chapter 9) that, in spite of lack of investment in nuclear power at the present time, for whatever the reason, a vigorous research and development programme is needed to ensure the acceptability of this energy source by the public when the need for nuclear power becomes paramount.

If coal is to become the only major source of energy from fossil fuel, it is obviously necessary to develop the most environmentally clean and efficient plant possible to use this fuel. The Working Party proposes that the R&D necessary to achieve this should be carried out. Carbon dioxide emission would be much reduced if satisfactory methods could be developed for removing and disposing of carbon dioxide in flue gases, both from coal burning plant and where appropriate from plant burning gas, refuse and biomass. The Working Party recommends R&D on this subject should start immediately as a matter of urgency. Finally, in its short list of longer term topics needing R&D the Working Party proposes new effort on alternative fuels and for methods of transport. A fall in the availability of oil will naturally stimulate the search for alternative fuels but it is essential that new transport fuels and new ways of using them meet the full range of restrictions on the emission of pollutants, including a limitation on the emission of carbon dioxide.

As stated in the opening paragraph of this Chapter, the main objective of the Working Party has been to put forward proposals that could be implemented immediately and which, if vigorously pursued, could lead to a significant fall in the build up of the greenhouse gases as the world moves into the next century. The recommendation is for a package of measures which, when taken together, should have the required effect if they can be applied effectively. The fact that they are all actions which are economically sensible and can be justified with regard to other potential benefits to the environment or to the standard of life gives hope that there will be less inertia against their realisation than might otherwise be the case.

The reasoning behind the choice of options made is given in Chapter 9 of this Report. The list of measures proposed is as follows:

- high efficiency natural gas-fired gas turbine combined cycle power plant
- natural gas-fired gas-turbine or gas engine combined heat and power in industry, the public and commercial sectors
- the promotion of wind energy
- combustion of non-recyclable waste, refuse and biomass with energy recovery where possible
- energy saving in public and commercial buildings and in the domestic sector. In particular:
  — thermal insulation and draught proofing
  — improved efficiency of space heating by choosing fuel type, controls and use of CHP
  — more efficient electrical appliances
  — low energy lighting and lighting controls
- encouraging the use of smaller, more efficient cars
- increasing the area of woodlands, especially combined with more use of wood, including as a fuel
- for listed CFC's;
  — adopt urgent measures to phase out their manufacture world-wide

- recover CFC's from plant and appliances so that they may be recycled or rendered harmless
- substitute substances with a much smaller greenhouse potential than the listed CFC's
• financial and technical support to the less developed countries of the world to help them
  - to formulate and implement global policies for forestry and agriculture
  - to help improve the efficiency of energy conservation and utilisation.

REFERENCES

DONALDSON, D. M. (Pers. Comm.) Private Communication to the Working Party 1990.
MELLANBY, K. (Pers. Comm.) Private Communication to the Working Party 1990.

THE WATT COMMITTEE ON ENERGY

Objectives, Historical Background and Current Programme

1. The objectives of the Watt Committee on Energy are:

   (a) to promote and assist research and development and other scientific or technological work concerning all aspects of energy;
   (b) to disseminate knowledge generally concerning energy;
   (c) to promote the formation of informed opinion on matters concerned with energy;
   (d) to encourage constructive analysis of questions concerning energy as an aid to strategic planning for the benefit of the public at large.

2. The concept of the Watt Committee as a channel for discussion of questions concerning energy in the professional institutions was suggested by Sir William Hawthorne in response to the energy price 'shocks' of 1973/74. The Watt Committee's first meeting was held in 1976, it became a company limited by guarantee in 1978 and a registered charity in 1980. The name 'Watt Committee' commemorates James Watt (1736–1819), the great pioneer of the steam engine and of the conversion of heat to power.

3. The members of the Watt Committee are 60 British professional institutions. It is run by an Executive on which all member institutions are represented on a rota basis. It is an independent voluntary body, and through its member institutions it represents some 500 000 professionally qualified people in a wide range of disciplines.

4. The following are the main aims of the Watt Committee:

   (a) to make practical use of the skills and knowledge available in the member institutions for the improvement of the human condition by means of the rational use of energy;
   (b) to study the winning, conversion, transmission and utilisation of energy, concentrating on the United Kingdom but recognising overseas implications;
   (c) to contribute to the formulation of national energy policies;
   (d) to identify particular topics for study and to appoint qualified persons to conduct such studies;
   (e) to organise conferences and meetings for discussion of matters concerning energy as a means of encouraging discussion by the member institutions and the public at large;
   (f) to publish reports on matter concerning energy;
   (g) to state the considered views of the Watt Committee on aspects of energy from time to time for promulgation to the member institutions, central and local government, commerce, industry and the general public as contributions to public debate;
   (h) to collaborate with member institutions and other bodies for the foregoing purposes both to avoid overlapping and to maximise cooperation.

5. Reports have been published on a number of topics of public interest. Notable among these are *The Rational Use of Energy* (an expression which the Watt Committee has always preferred to 'energy conservation' or 'energy efficiency'), *Passive Solar Energy, Air Pollution, Acid Rain and the Environment, The Chernobyl Accident and its Implications for the United Kingdom, The Membrane Alternative* and *Renewable Energy Sources*. Others are in preparation.

6. Those who serve on the Executive, working groups and sub-committees or who contribute in any way to the Watt Committee's activities do so in their independent personal capacities without remuneration to assist with these objectives.

7. The Watt Committee's activities are coordinated by a small permanent secretariat. Its income is generated

by its activities and supplemented by grants by public, charitable, industrial and commercial sponsors.
8. The latest Annual Report and a copy of the Memorandum and Articles of Association of the Watt Committee on Energy may be obtained on application to the Secretary.

Enquiries to:
The Information Officer,
The Watt Committee on Energy,
Savoy Hill House,
Savoy Hill, London WC2R 0BU
Telephone: 071-379 6875

Member Institutions of The Watt Committee on Energy

British Association for the Advancement of Science
*British Nuclear Energy Society
British Wind Energy Association
Chartered Institute of Building
Chartered Institute of Building Services Engineers
*Chartered Institute of Management Accountants
*Chartered Institute of Transport
Combustion Institute (British Section)
Geological Society of London
Hotel Catering and Institutional Management Association
Institute of Biology
Institute of British Foundrymen
Institute of Ceramics
Institute of Chartered Foresters
*Institute of Energy
Institute of Home Economics
Institute of Hospital Engineering
Institute of Internal Auditors (United Kingdom Chapter)
Institute of Management Services
Institute of Marine Engineers
Institute of Mathematics and its Applications
Institute of Metals
*Institute of Petroleum
Institute of Physics
Institute of Purchasing and Supply
Institute of Refrigeration
Institute of Wastes Management
Institution of Agricultural Engineers
*Institution of Chemical Engineers
*Institution of Civil Engineers
Institution of Electrical and Electronics Incorporated Engineers
*Institution of Electrical Engineers
Institution of Engineering Designers
*Institution of Gas Engineers
Institution of Geologists
*Institution of Mechanical Engineers
Institution of Mining and Metallurgy
Institution of Mining Engineers
Institution of Nuclear Engineers
*Institution of Plant Engineers
Institution of Production Engineers
Institution of Structural Engineers
International Solar Energy Society—UK Section
Operation Research Society
Plastics and Rubber Institute
Royal Aeronautical Society
Royal Geographical Society
*Royal Institute of British Architects
Royal Institution
Royal Institution of Chartered Surveyors
Royal Institution of Naval Architects
Royal Meteorological Society
Royal Society of Arts
*Royal Society of Chemistry
Royal Society of Health
Royal Town Planning Institute
*Society of Business Economists
Society of Chemical Industry
Society of Dyers and Colourists
Textile Institute

* Denotes permanent members of The Watt Committee Executive

Watt Committee Reports

2. Deployment of National Resources in the Provision of Energy in the UK
3. The Rational Use of Energy
4. Energy Development and Land in the United Kingdom
5. Energy from the Biomass
6. Evaluation of Energy Use
7. Towards an Energy Policy for Transport
8. Energy Education Requirements and Availability
9. Assessment of Energy Resources
10. Factors Determining Energy Costs and an Introduction to the Influence of Electronics
11. The European Energy Scene
13. Nuclear Energy: a Professional Assessment
14. Acid Rain
15. Small-Scale Hydro-Power
17. Passive Solar Energy in Buildings
18. Air Pollution, Acid Rain and the Environment
19. The Chernobyl Accident and its Implications for the United Kingdom
20. Gasification: Its Role in the Future Technological and Economic Development of the United Kingdom
21. The Membrane Alternative: Energy Implications for Industry
22. Renewable Energy Sources

For further information and to place orders, please write to:

ELSEVIER SCIENCE PUBLISHERS
Crown House, Linton Road, Barking, Essex IG11 8JU, UK

Customers in North America should write to:

ELSEVIER SCIENCE PUBLISHING CO., INC.
655 Avenue of the Americas, New York, NY 10010, USA

INDEX

Acid deposition 5
Aerosol propellants 11
Agriculture 49-51
  developing countries 67-8
  energy consumption 24
  main sources of greenhouse
    gases 62
Air conditioning 11
  potential for saving energy 44
Air transport, energy consumption 23
Air travel 54
  energy costs 59
Airless drying 46
Alcohol fuels 57

Biofuels 31-2
Biomass 49, 57, 62, 63

CAFE (corporate average fuel
  efficiency) regulations 58
Car replacement 54-5
Carbon cycling 6-8
Carbon dioxide emissions 1, 2,
  19, 54, 65, 73, 79
  atmospheric concentration 6-9
  atmospheric response to 8-9
  by direct use and fuel type 20
  by end user and fuel type 21
  coal burning plant 88
  contribution to greenhouse
    effect 13
  developed world 65
  developing countries 65-6
  domestic sector 36
  effect of atmospheric lifetimes 13
  energy rise and 45
  forestry 62-3, 68-70
  fossil fuel combustion 27
  framework of reducing 78
  gas fuelled vehicles 58
  global 17
  government policy 58
  1950-88 21
  overall reductions from
    dwellings 39-40
  projections of future 8-10
  public transport 59
  removal from power plant or
    industrial flue gases 30
  road transport sector 55
  sources 5-6
  Spain 82-3
  stabilisation 8-9
Carbon emissions by sector and
  fuel type 79
Catering, energy usage 43
Central Electricity Generating
  Board (CEGB) 27, 30
Chemicals industry, improving
  energy efficiency 48-9
Chlorofluorocarbons (CFCs) 11-12,
  61-2, 73-4, 89
  contribution to greenhouse
    effect 13
  estimates of production and
    emission 19
  global emissions 19
  replacements for 12, 61
Circulation models 2
Climate change 3
Climate modelling 2-3
Climate response 2
Clubs, energy usage 43
Coal burning plant, carbon
  dioxide emissions 88
Combined heat and power (CHP) 29,
  41, 43, 49, 74, 81
  large scale 46
  small scale 46
Commercial sector, energy usage
  41-3
Company cars 59
Compressed natural gas (CNG) 57
Condensing boilers 46

Condensing combined cycle 49
Cooking
    improvements in 39
    potential for saving energy 44-5
Coppicing 32

Defence establishments, energy usage 43
Dehumidification, potential for saving energy 45
Developing countries 65-6
    agriculture 67-8
    energy sector 66-7
    new power and steam raising plant 67
    use of existing plant 66-7
Distribution, energy usage 43
Domestic appliances, improvements in 39
Domestic sector, possible future emissions 41
Draught proofing 37

East Germany 75
Eastern Europe 75
Education, energy usage 44
Electric cars 58
Electrical applicances, efficiency of 46
Electricity
    automotive applications 58
    production in France 79
Electricity generating plant, thermal efficiency 27-8
Electrolysis of water 33
Energy and Technology Support Unit (ETSU) 41
Energy consumption
    agriculture 24
    air transport 23
    and world population 9
    by end use 42
    by fuel type 42
    by source 79
    carbon dioxide emission 45
    conversion to carbon dioxide emissions 19
    domestic sector 24, 35-41
    iron and steel industry 22-3
    projected 9
    public administration 23
    rail transport 24
    road transport 23
    trends in 22-4
    water transport 24
Energy conversion 74
    and release of greenhouse gases 27-34
    France 80
Energy cost-saving 48
Energy demand 87-8
    recent trends 44
    UK primary 19
Energy efficiency 44
    cost-effectiveness 36-7
    increases in 36
    new technologies 45
    technology investment 78
Energy forestry 32
Energy savings 74
    cost effective 38
    France 81
    technically possible 38
Energy sector, developing countries 66-7
Energy usage
    commercial sector 41-3
    effect on release of greenhouse gases 35-51
    monitoring and targeting 48
    public buildings sector 41, 43-4
    transportation 53-9
Engine technology 55-6
Equilibrium experiments 2
Ethanol 57
European Centrally Planned Economies (CPE's) 17
European Communities 84-6

Feedback processes 2
Foam blowing 11
Food industry 49-51
    primary energy involved in 50
Forestry 62-3, 68-70, 74
    France 81
Fossil-fired technology 28-9
Fossil fuels
    hydrogen from 33
    increased use of 65
Freons 61
Fuels
    alternative 57
    substitution of 29
    UK growth in 54

Gas fuelled vehicles, carbon dioxide emissions 58
Gas fuels 57-8
Geothermal power 31
German perspective 78-9
Global warming 77, 87
Greenhouse effect
    definition 1
    natural 1
    scientific assessment of 77

Index

Greenhouse gases 1, 2, 3, 73
  and energy conversion 27-34
  assessing future importance of 5-15
  fraction of emissions remaining in the atmosphere 13
  France 79
  global emissions 17-19
  lifetimes in atmosphere 12
  non-energy related sources and sinks 61-3
  relative importance of emissions of 12-14
  UK emissions 19
  warming potential of 12
  see also under specific gases

HCFC 61-2
Health premises, energy usage 44
Heat balance 1-2
Heat generation, alternative systems 40-1
Heat pumps 46
Heat recovery 49
Heating appliance efficiencies, improvements in 39
HFC 61
High temperature heating/melting 49
Hybrid vehicles 58
HYDRO CARB process 33
Hydrocarbons 62
Hydrogen 32-3
  production methods 33
  storage 58
Hydrogen vehicles 58
Hydropower generation 32

Ice core measurements 1
Industry sector 46-9
  energy efficiency literature 47
  energy efficient technology 49
Infra-red absorption strength 12
Insulation 37
Iron and steel industry, energy consumption 22-3

Lighting
  improvements in 39
  potential for saving energy 44
Liquefied hydrogen 32-3
Liquefied petroleum gas (LPG) 57
Local government, energy usage 43

Methane 1, 62, 68, 73, 80
  atmospheric concentration of 10
  contribution to greenhouse effect 13
  effect of atmospheric lifetimes 13
  emissions 10, 74
  future concentrations 10
  global budget 10
  global emissions 18
  UK emissions 22
Methanol 57

National government, energy usage 43
Natural gas 74
Natural gas fired gas-turbines 74
New technologies, energy efficiency 45
Nitrogen oxides 5, 11
Nitrous oxide 11, 80
  contribution to greenhouse effect 13
  global emissions 18
Nuclear energy 30
Nuclear industry 88

Oil price shocks 53
Ozone 11
  stratospheric 11

Paddy fields 62
Photocatalysis 32
Photolytic dissociation of water 33
Photovoltaic cells 32
Phytoplankton, life cycles of 8
Policy formulation implications 73-6
Power generation, alternative systems 40-1
Private transport 53
Public administration, energy consumption 23
Public buildings sector, energy usage 41, 43-4
Public houses, energy usage 43
Public transport
  carbon dioxide emissions 59
  government policy 59

Rail transport 59
  energy consumption 24
Refrigerants 11
Refrigeration, potential for saving energy 45
Refuse derived fuel (RDF) 31
Regulation 73
Renewable energy 30-2, 46, 74
Residential accommodation, energy usage 43
Retail services, energy usage 43
Rice fields 62, 68
Road congestion crisis 54
Road haulage sector 53

Road traffic growth 54
Road transport
  carbon dioxide emissions 55
  energy consumption 23
Ruminant farm animals 62

Satellite observations 1
Shops, energy usage 43
Short-list of longer term
  research and development topics 75
Short-list of short-term measures 75
Solar energy 32, 49, 74
Solar heating 44, 46
Solar radiation 1, 46
Solvents 62
Space heating, potential for
  saving energy 44
Spain, environmental concerns 81-4
Steam raising plant 49
Stratospheric ozone depletion 11
Sulphur dioxide 5, 11

Taxation 73
Technology innovation/plant
  replacement 48
Thermal dissociation of water 33
Tidal power 30-1
Traffic management 56
Transportation 74
  energy usage 53-9
  France 81
  possible future strategy 55
  see also Public transport; Road
    transport
Tropospheric ozone 11

Vegetarianism 68
Vehicle efficiency improvements 56
Vehicle emissions, planning
  regulations 58
Vehicle ownership 53
Vehicle survival rates 54

Waste utilisation 31-2
Water
  electrolysis 33
  photolytic dissociation 33
  thermal dissociation 33
Water borne transport 59
Water heating, potential for
  saving energy 44
Water transport, energy
  consumption 24
Water vapour 1, 2
Watt Committee on Energy 91-4
Wave power 31
Wind power 32, 74
World Climatic Programme 55
World Commission on Environment
  and Development 9
World population, and energy
  consumption 9
World Wide Fund for Nature 55